Creating Rhythms

Stefan Hollos and J. Richard Hollos

Creating Rhythms
by Stefan Hollos and J. Richard Hollos
Paper ISBN 978-1-887187-22-0
Ebook ISBN 978-1-887187-23-7

Abrazol Publishing

an imprint of Exstrom Laboratories LLC
662 Nelson Park Drive, Longmont, CO 80503-7674 U.S.A.

About the Cover

A Christoffel word playing drums. Cover design made with the help of Inkscape.

Contents

WHAT THIS BOOK IS ABOUT

This book is about creating rhythms via a mathematical approach that uses combinatorics, de Bruijn sequences, Christoffel words, paper folding, and probability. We don't expect the reader to know what any of these terms mean. They are all explained in the book.

The methods discussed in the book can be used to produce an almost endless variety of new rhythms, and in the process, traditional ones pop up too.

Other than the last chapter on stochastic rhythms, the mathematics used is at a very elementary level. If you're comfortable with a little bit of algebra and you know for example that $4! = 1 \cdot 2 \cdot 3 \cdot 4 = 24$ then you should have no problem.

On the book's website (abrazol.com/books/rhythm1/), you will find command-line programs for doing calculations and creating rhythms. The programs are written in the C programming language, and will have to be compiled before you can use them. You do not have to know C to use the programs or understand the contents of the book. There is a C language compiler for every major operating system. A good one that is also free is gcc. There is nothing operating system specific about any of the programs, so you should have no problems compiling them on any computer.

The book contains many rhythms, and those that refer to the book's website have MIDI files associated with them that you can listen to. You can go to the website and listen to all the rhythms, or if you're reading an ebook version, then there are individual links to the MIDI files. There is a MIDI player for every major operating system. A good free one is TiMidity.

May you create beautiful rhythms.

Stefan Hollos and Richard Hollos
Exstrom Laboratories LLC
Longmont, Colorado
Feb 21, 2014

INTRODUCTION

If you look at rhythms as any recurring sequence of events, then we are immersed in them: days and nights, tides, the lunar cycle, the seasons, harvests, migrations, the solar cycle. So we and the other animals have an intimate familiarity with, and often an affinity for them, even enjoying their representation as a sequence of sounds.

The most basic rhythm making instrument is the drum. Gorillas use their chests as drums, and macaque monkeys show social dominance by drumming objects. One of the most primitive man-made drums, from the Brazilian Catuquinaru Indians, was made by digging a hole in the ground, then dragging a hollow log over it. [1]

Clearly one of the major uses of rhythms in the animal world is for communication and signaling. But rhythms are also the foundation of music so it is natural to ask what it is about a particular rhythm that makes it interesting.

For something to be interesting to us it needs to have structure at many levels. A tree has branching patterns at different levels. There is the branching of the main trunk and those branches will further branch and so on, down to the branching of the twigs. It is the way this

[1] The Drum: a History, pg 90, Matt Dean, 2012

hierarchy of branching produces the whole tree that makes it interesting.

Mountains are another example. If mountains were simply a conical shaped part of the landscape, that you could climb up, then they would not be nearly as interesting as they are. Anyone who has climbed mountains knows that they can be whole worlds unto themselves. There are hidden valleys, ridges, cliffs, small plateaus, small lakes, and countless other features that are washed out when looking at the mountain from a distance. It is the variety of features at different levels that makes mountains so fascinating.

There are various ways that levels of structure can be produced. Self similarity is one way. In this case the whole is composed of scaled copies of itself. It you look at one of the main branches of a tree, chances are that it will look similar to the tree as a whole. A mountain is the sum of smaller mountains. A coastline looks the same at different scales. The study of self similarity is called fractal geometry and the person most responsible for developing the field is Benoit Mandelbrot. It is still very much an active area of research.

One way to get self similarity is to have simultaneous correlations at many different levels. We say two things are correlated if one of them depends in some way on the other. The position of a twig depends on the position of the branch it is connected to, which in turn is dependent on the position of the branch it is connected to, and so on down to the position of the tree itself.

Two of the methods discussed in this book are ways to directly create rhythms with a self similar structure. We call them Christoffel rhythms and Folding rhythms. Both these methods can be used to create an almost infinite variety of rhythms, either individually or combined in different ways. Some of the most popular rhythms in the world turn out to be Christoffel rhythms.

We start by looking at general combinatorial methods for creating rhythms. These are based on what are called integer compositions, binary necklaces, integer partitions, and de Bruijn sequences. Next come Christoffel rhythms, Folding rhythms, Natural rhythms and finally stochastic rhythms which come from generating random numbers with different degrees of correlation, i.e. different degrees of predictability.

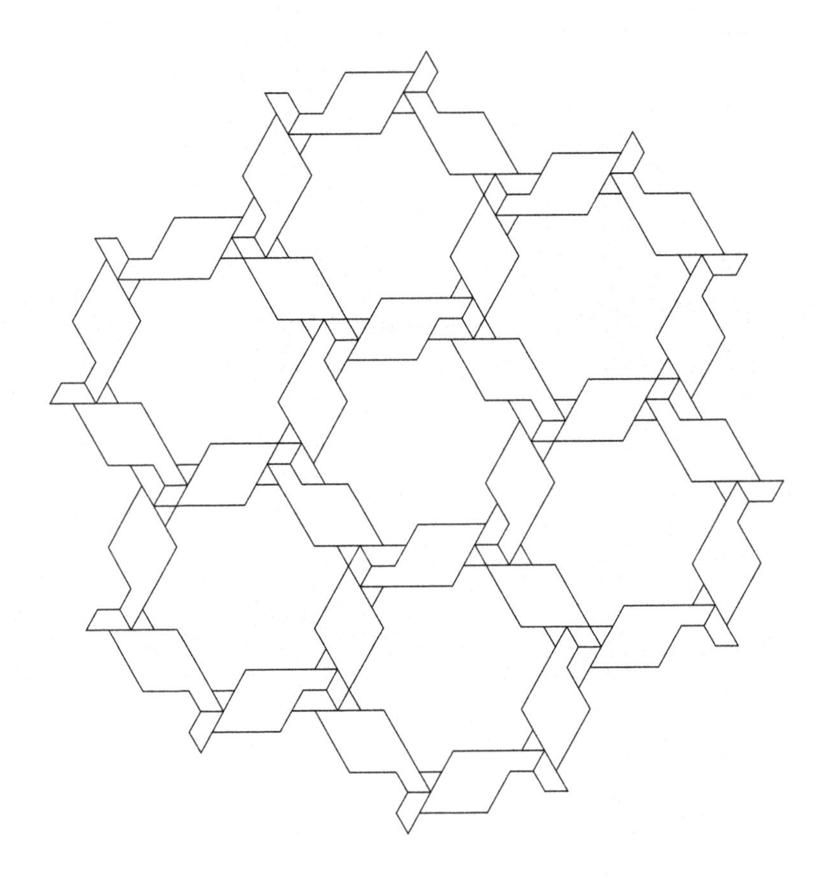

WHAT IS A RHYTHM?

In its most general form, a rhythm is simply a recurring sequence of events. The motion of the sun, the planets, and the moon across the sky can produce rhythms. Even though the individual motions have more or less constant periodicities, together they can produce complex patterns. The orbit of the moon and the rotation of the earth interact to produce the rhythm of the tides. The orbit of the earth around the sun produces the rhythm of the seasons. The sun has an eleven year sunspot cycle that varies the intensity of colorful auroras in the night sky.

The beating of a heart produces complex rhythms. Here you have a recurring event but the timing can vary widely, speeding up, slowing down, and sometimes missing a beat altogether. Walking is a rhythmic process where the timing and nature of the rhythm can change suddenly when going from a walk to a run. These are examples of free rhythms where there are recurring events but the timing can be almost anything.

In this book we are going to look at mostly measured rhythms where time is divided into fixed intervals and events always occur at the beginning of an interval. Two events are always separated in time by an integer number of intervals. Most music uses measured rhythms where the events are note onsets. By a note we mean any sound that can be produced by a musical instrument. A note can be produced by striking a drum, pushing a key on a piano, or strumming a chord on a guitar. A note can also sometimes mean a silence which is called a rest in music.

In music the time intervals of a measured rhythm are called pulses or beats. We will call them pulses so that we can reserve the word beat for drum beats. A rhythm consists of a sequence of note onsets that are separated by integer numbers of pulses. The length of each note is also an integer number of pulses. Two rhythms may be played simultaneously to create a more complex rhythm. We call two or more rhythms played simultaneously a rhythm set. A rhythm is usually repeated several times, or two or more rhythms may be played in a repeating and alternating sequence.

MEASURED RHYTHMS

When we talk about rhythms we usually have in mind some fixed number of pulses containing a fixed number of note onsets. We denote the number of pulses as n and the number of note onsets as m. Each note is composed of at least one pulse and each pulse has no more than one note therefore m must be less than or equal to n. Rhythms with multiple notes on a pulse can be represented as multiple rhythms played simultaneously. For example the beats of a bass drum and a snare drum are represented as separate rhythms played simultaneously.

There are many ways to represent rhythms. The simplest representation is as a string of 1's and 0's where each character corresponds to a pulse. A pulse with a note onset is represented by a 1. A pulse with no note onset is represented by a 0. We call this a binary string representation.

Another representation we will use is called box notation. Here a pulse with a note onset is represented by a ■ and one with no note onset is represented by a □.

For example the binary string 10100 represents a rhythm with 5 pulses and 2 notes. Counting the pulses starting at 0, the first note is at pulse 0 and the second note is at pulse 2. In box notation the rhythm is ■□■□□.

We are not going to be concerned with the length of notes. Since we are only looking at rhythms, it is only the note onsets that are important to us. You can assume that a note may last from a single pulse up to the number of pulses until the next note onset.

Another way to represent rhythms is in terms of intervals between note onsets. For the rhythm 10100 the interval notation is just 2 3 which means the first note gets 2 pulses and the second gets 3.

In the interval notation, the sum of all the intervals equals the total number of pulses in the rhythm. So you can look at the construction of a rhythm as a process by which n pulses are distributed among m notes such that each note gets at least one pulse. This is analogous to dividing the number n into m parts so that each part is greater than or equal to 1. In mathematics such a division is called a composition of n into m parts and there is a relatively simple formula for the number of ways it can be done. The formula is:

$$\binom{n-1}{m-1} = \frac{(n-1)!}{(m-1)!(n-m)!}$$

The formula is called a binomial coefficient and most scientific calculators, spreadsheets and mathematical software packages will be able to evaluate it. If the formula looks totally alien to you don't worry, it's really just for reference purposes, in case you want to know exactly how many rhythms with n pulses and m note onsets there are.

Suppose for example that we have 2 notes and 4 pulses, then the formula says there are 3 possible compositions or rhythms. They are: $3+1$, $1+3$, and $2+2$. In the first composition, the first note gets 3 pulses and the second only 1. In the second composition this is reversed and in the third each note gets 2 pulses. Note that if each note has a common multiple number of pulses then the pulse length could be redefined to the common multiple so that $2+2$ becomes just $1+1$. The table below shows the number of compositions for rhythms of length 4 through 9 and from a single note to a note on every pulse.

m \ n	4	5	6	7	8	9
1	1	1	1	1	1	1
2	3	4	5	6	7	8
3	3	6	10	15	21	28
4	1	4	10	20	35	56
5	0	1	5	15	35	70
6	0	0	1	6	21	56
7	0	0	0	1	7	28
8	0	0	0	0	1	8
9	0	0	0	0	0	1

Table 1: Number of compositions of n into m parts.

UNRESTRICTED RHYTHMS

What we have described so far are unrestricted rhythms where there are no restrictions on the size of intervals between note onsets. The only requirement is that m notes occur within the n pulses of the rhythm and that the first pulse is a note. These are called unrestricted compositions of the integer n into m parts.

As an example of unrestricted compositions, tables 2, 3, 4, and 5 below show all possible compositions of $n = 4, 5, 6, 7$ and $m = 3$ along with binary and box representations of the corresponding rhythm.

The program `compm.c` available on the book's website will create all compositions for a given n and m. Be warned that this can be a very large number. For example the number of compositions of 32 pulses and 16 notes is 300540195. This is more than 300 million possible rhythms. Of course not all of them will sound interesting. To more systematically look for interesting rhythms, we need to be able to put restrictions on the size of the intervals between notes. This is the subject of the next section.

Intervals	Binary	Box
1 1 2	1110	■■■□
1 2 1	1101	■■□■
2 1 1	1011	■□■■

Table 2: All 4 pulse rhythms with 3 onsets: `compm 4 3`

Intervals	Binary	Box
1 1 3	11100	■■■□□
1 2 2	11010	■■□■□
1 3 1	11001	■■□□■
2 1 2	10110	■□■■□
2 2 1	10101	■□■□■
3 1 1	10011	■□□■■

Table 3: All 5 pulse rhythms with 3 onsets: `compm 5 3`

Intervals	Binary	Box
1 1 4	111000	■■■□□□
1 2 3	110100	■■□■□□
1 3 2	110010	■■□□■□
1 4 1	110001	■■□□□■
2 1 3	101100	■□■■□□
2 2 2	101010	■□■□■□
2 3 1	101001	■□■□□■
3 1 2	100110	■□□■■□
3 2 1	100101	■□□■□■
4 1 1	100011	■□□□■■

Table 4: All 6 pulse rhythms with 3 onsets: `compm 6 3`

Intervals	Binary	Box
1 1 5	1110000	■■■□□□□
1 2 4	1101000	■■□■□□□
1 3 3	1100100	■■□□■□□
1 4 2	1100010	■■□□□■□
1 5 1	1100001	■■□□□□■
2 1 4	1011000	■□■■□□□
2 2 3	1010100	■□■□■□□
2 3 2	1010010	■□■□□■□
2 4 1	1010001	■□■□□□■
3 1 3	1001100	■□□■■□□
3 2 2	1001010	■□□■□■□
3 3 1	1001001	■□□■□□■
4 1 2	1000110	■□□□■■□
4 2 1	1000101	■□□□■□■
5 1 1	1000011	■□□□□■■

Table 5: All 7 pulse rhythms with 3 onsets: `compm 7 3`

RESTRICTED RHYTHMS

Besides specifying the number of notes m occurring within the n pulses, we can also impose restrictions on the size of the intervals between note onsets. The program `compam.c` will generate all the compositions of n into m parts, with sizes of parts restricted to a set of allowed sizes.

Table 6 shows all the compositions of 16 into 5 parts with sizes of parts restricted to 2, 3, or 4, generated by command:
`compam 16 5 2 3 4`
The box rhythm notation is also shown. There are a total of 45 of these rhythms. Without restriction on part sizes there would be 1365, so we've gotten rid of the vast majority of possible rhythms with 5 onsets by restricting size of onset intervals to 2, 3, or 4. This table contains five of the world's most popular rhythms: Son Clave (3 3 4 2 4), Rumba Clave (3 4 3 2 4), Bossa Nova (3 3 4 3 3), Gahu (3 3 4 4 2), and Shiko (4 2 4 2 4). You can listen to them, as well as many other traditional rhythms, on the book's website.

If you add 1 to the set of allowed onset intervals, then you get the 20 additional rhythms shown in table 7. If you also add 5 to the set of allowed onset intervals, then you get a total of 365 possible 16 pulse rhythms with 5 note onsets, including the popular Soukous (3 3

add 1 to Allowed = +20 more

add 5 to Allowed = 365

4 1 5). This gives you an entire calendar of rhythms, one for each day of the year.

An interesting thing to do is play a rhythm from this list together with a cyclic shift of itself or play it together with one of its neighbors. For example, if you take the Soukous rhythm and create one rhythm pair for each possible cyclic shift with itself, a couple interesting sounding ones are shown below, and can be listened to on the book's website, using low tom for the original rhythm and high tom for the shifted one.

Soukous and its right shift by 2

■□□■□□■□□□■■□□□□
□□■□□■□□■□□□■■□□

Soukous and its right shift by 10

■□□■□□■□□□■■□□□□
■□□□■■□□□□■□□■□□

Doing the same for the other five popular rhythms (Son Clave, Rumba Clave, Bossa Nova, Gahu, Shiko), some interesting sounding ones are listed below.

Son Clave and its right shift by 14

■□□■□□■□□□■□■□□□
□■□□■□□□■□■□□□■□

Rumba Clave and its right shift by 5

■□□■□□□■□□■□■□□□
□■□□□■□□■□□□■□□■

Intervals	Box	
2 2 4 4 4	■□■□■□□□■□□□■□□□	
2 3 3 4 4	■□■□□■□□■□□□■□□□	
2 3 4 3 4	■□■□□■□□□■□□■□□□	
2 3 4 4 3	■□■□□■□□□■□□□■□□	
2 4 2 4 4	■□■□□□■□■□□□■□□□	
2 4 3 3 4	■□■□□□■□□■□□■□□□	
2 4 3 4 3	■□■□□□■□□■□□□■□□	
2 4 4 2 4	■□■□□□■□□□■□■□□□	
2 4 4 3 3	■□■□□□■□□□■□□■□□	
2 4 4 4 2	■□■□□□■□□□■□□□■□	
3 2 3 4 4	■□□■□■□□■□□□■□□□	
3 2 4 3 4	■□□■□■□□□■□□■□□□	
3 2 4 4 3	■□□■□■□□□■□□□■□□	
3 3 2 4 4	■□□■□□■□■□□□■□□□	
3 3 3 3 4	■□□■□□■□□■□□■□□□	
3 3 3 4 3	■□□■□□■□□■□□□■□□	
3 3 4 2 4	■□□■□□■□□□■□■□□□	Son Clave
3 3 4 3 3	■□□■□□■□□□■□□■□□	Bossa Nova
3 3 4 4 2	■□□■□□■□□□■□□□■□	Gahu
3 4 2 3 4	■□□■□□□■□■□□■□□□	
3 4 2 4 3	■□□■□□□■□■□□□■□□	
3 4 3 2 4	■□□■□□□■□□■□■□□□	Rumba Clave
3 4 3 3 3	■□□■□□□■□□■□□■□□	
3 4 3 4 2	■□□■□□□■□□■□□□■□	
3 4 4 2 3	■□□■□□□■□□□■□■□□	
3 4 4 3 2	■□□■□□□■□□□■□□■□	
4 2 2 4 4	■□□□■□■□■□□□■□□□	
4 2 3 3 4	■□□□■□■□□■□□■□□□	
4 2 3 4 3	■□□□■□■□□■□□□■□□	
4 2 4 2 4	■□□□■□■□□□■□■□□□	Shiko
4 2 4 3 3	■□□□■□■□□□■□□■□□	
4 2 4 4 2	■□□□■□■□□□■□□□■□	
4 3 2 3 4	■□□□■□□■□■□□■□□□	
4 3 2 4 3	■□□□■□□■□■□□□■□□	
4 3 3 2 4	■□□□■□□■□□■□■□□□	
4 3 3 3 3	■□□□■□□■□□■□□■□□	
4 3 3 4 2	■□□□■□□■□□■□□□■□	
4 3 4 2 3	■□□□■□□■□□□■□■□□	
4 3 4 3 2	■□□□■□□■□□□■□□■□	
4 4 2 2 4	■□□□■□□□■□■□■□□□	
4 4 2 3 3	■□□□■□□□■□■□□■□□	
4 4 2 4 2	■□□□■□□□■□■□□□■□	
4 4 3 2 3	■□□□■□□□■□□■□■□□	
4 4 3 3 2	■□□□■□□□■□□■□□■□	
4 4 4 2 2	■□□□■□□□■□□□■□■□	

Table 6: Rhythms with intervals restricted to 2 3 4

all have 5 onsets

Intervals	Box
1 3 4 4 4	■■□□■□□□■□□□■□□□
1 4 3 4 4	■■□□□■□□■□□□■□□□
1 4 4 3 4	■■□□□■□□□■□□■□□□
1 4 4 4 3	■■□□□■□□□■□□□■□□
3 1 4 4 4	■□□■■□□□■□□□■□□□
3 4 1 4 4	■□□■□□□■■□□□■□□□
3 4 4 1 4	■□□■□□□■□□□■■□□□
3 4 4 4 1	■□□■□□□■□□□■□□□■
4 1 3 4 4	■□□□■■□□■□□□■□□□
4 1 4 3 4	■□□□■■□□□■□□■□□□
4 1 4 4 3	■□□□■■□□□■□□□■□□
4 3 1 4 4	■□□□■□□■■□□□■□□□
4 3 4 1 4	■□□□■□□■□□□■■□□□
4 3 4 4 1	■□□□■□□■□□□■□□□■
4 4 1 3 4	■□□□■□□□■■□□■□□□
4 4 1 4 3	■□□□■□□□■■□□□■□□
4 4 3 1 4	■□□□■□□□■□□■■□□□
4 4 3 4 1	■□□□■□□□■□□■□□□■
4 4 4 1 3	■□□□■□□□■□□□■■□□
4 4 4 3 1	■□□□■□□□■□□□■□□■

Table 7: Additional rhythms with 1 as an onset interval

Bossa Nova and its right shift by 8

■□□■□□■□□□■□□■□□

□□■□□■□□■□□■□□■□

Gahu and its right shift by 14

■□□■□□■□□□■□□□■□

□■□□■□□□■□□□■□■□

Shiko and its right shift by 9

■□□□■□■□□□■□■□□□

□□□■□■□□□■□□□■□■

The number of ways to play these rhythms is almost endless.

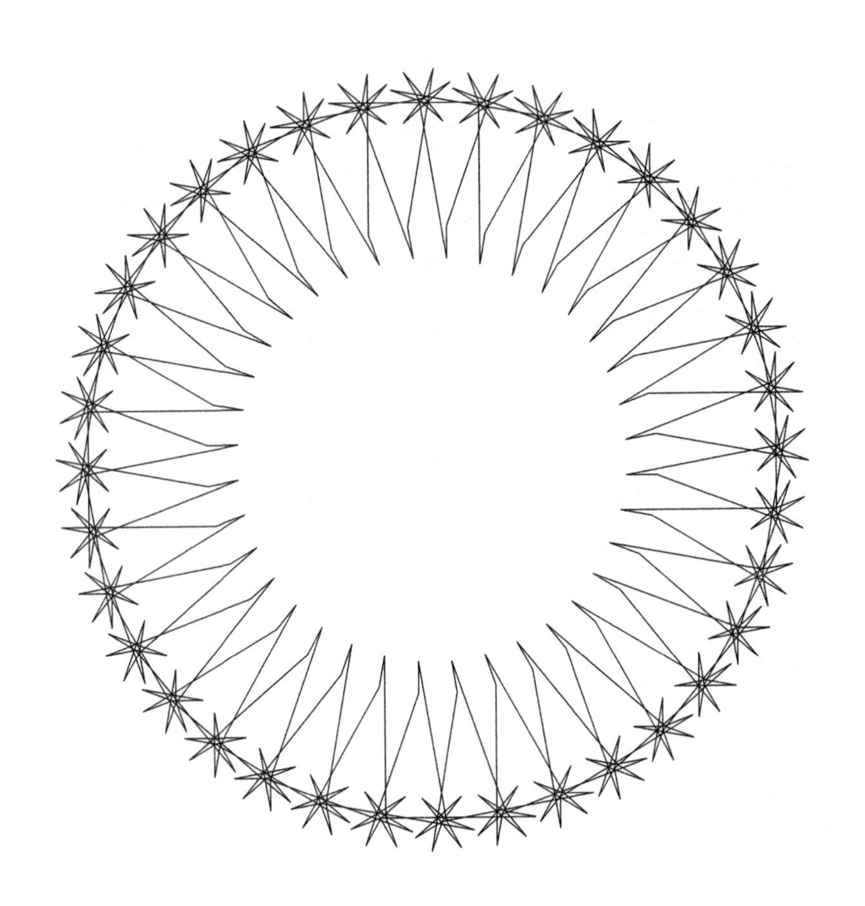

RHYTHM NECKLACES

Looking at the tables for the 16 pulse rhythms, you will notice many of them are simply cyclic shifts of others. For example, the first rhythm in table 6 is 2 2 4 4 4. If you do a cyclic shift of these intervals to the left, then you get the rhythm 2 4 4 4 2 which also appears in the table as the tenth entry.

Sometimes you may want to generate a set of rhythms where no rhythm is a cyclic shift of another. The rhythms in such a set are called rhythm necklaces. The program `neckam.c` is just like the program `compam.c` except that the rhythms it produces are not cyclically related to each other. For example the 16 pulse necklaces with 5 onsets, and intervals restricted to 2, 3, and 4 are shown in table 8. There are 9 of them.

How many rhythm necklaces with n pulses and m notes are there? The formula for this is not as simple as the one for compositions. You probably will never need to evaluate the formula, but we are going to present it as a reference. If you really need to know how many necklaces there are for reasonable values of n and m, you can just generate them with the program `neckm.c` and count them, or you can look it up in the table below. If you're interested in large values of n and m, then you may have to evaluate the formula, so here it is.

Intervals	Box
2 2 4 4 4	■□■□■□□□■□□□■□□□
2 3 3 4 4	■□■□□■□□■□□□■□□□
2 3 4 3 4	■□■□□■□□□■□□■□□□
2 3 4 4 3	■□■□□■□□□■□□□■□□
2 4 2 4 4	■□■□□□■□■□□□■□□□
2 4 3 3 4	■□■□□□■□□■□□■□□□
2 4 3 4 3	■□■□□□■□□■□□□■□□
2 4 4 3 3	■□■□□□■□□□■□□■□□
3 3 3 3 4	■□□■□□■□□■□□■□□□

Table 8: 16 pulse necklaces with 5 onsets, and intervals restricted to 2, 3, and 4

$$N(n,m) = \frac{1}{n} \sum_{d|n,m} \phi(d) \binom{n/d}{m/d}$$

The sum is over all the divisors, d, of both n and m, and the function $\phi(d)$ is called Euler's totient function. The value of $\phi(d)$ is equal to the number of integers less than d that are relatively prime to d. You can probably evaluate the function using a mathematical software package such as Mathematica, Maple, or a free open source equivalent called Maxima. Table 9 below shows the values for $n = 8, 9, \ldots, 16$ and $m = 3, 4, 5, 6$.

You can generate all the n pulse rhythm necklaces with m notes using the program `neckm.c`. If you want to limit the size of the intervals, then you can generate the necklaces using the program `neckam.c`. All the rhythm necklaces generated with these programs are unique with respect to cyclic shifts, i.e. you cannot do a cyclic shift on one of them and get another from the set.

		m			
		3	4	5	6
	8	7	10	7	4
n	9	10	14	14	10
	10	12	22	26	22
	11	15	30	42	42
	12	19	43	66	80
	13	22	55	99	132
	14	26	73	143	217
	15	31	91	201	335
	16	35	116	273	504

Table 9: Number of rhythm necklaces with n pulses and m notes.

RHYTHMS FROM PARTITIONS

There are many other ways to restrict the rhythms that go into a set. For example, two rhythms may be related by a reversal and cyclic shift of the intervals. If you reverse the rhythm 2 3 4 3 4 in table 8 you get 4 3 4 3 2 and if you do a cyclic shift to the right by one you get 2 4 3 4 3, which is also in the table. The other pair of rhythms in table 8 that are related like this is 2 3 3 4 4 and 2 4 4 3 3. If you keep only one from each pair then you are left with 7 rhythms instead of 9.

Another restriction is to eliminate rhythms that are repetitions of a shorter rhythm. For example the set of all 16 pulse rhythms with 6 onsets and intervals of size 2, 3 or 4 contains the rhythm 2 2 4 2 2 4 which is just a repetition of the shorter rhythm 2 2 4.

This idea of restricting rhythms is a top down view. You can also take a bottom up view by starting with all possible sets of intervals for a given number of pulses and building the rhythms from them. In mathematics these sets of intervals are called partitions of n, the number of pulses. Partitions of n into m parts are not distinguished by the order of the parts. For example 2 3 4 is a partition of $9 = 2 + 3 + 4$ into 3 parts. Any permutation of the numbers 2, 3, and 4, such as 3 2 4 is the same partition.

Let's look at 16 pulse rhythms with 5 onsets and intervals restricted to 2, 3, or 4. The 45 possible rhythms are shown in table 6. The partitions of 16 into 5 parts with part sizes restricted to 2, 3, or 4 are: 2 2 4 4 4, 2 3 3 4 4, and 3 3 3 3 4. All the 45 rhythms in table 6 are permutations of the parts in one of these partitions. There are 10 ways to permute the parts in 2 2 4 4 4, 30 ways to permute the parts in 2 3 3 4 4, and 5 ways to permute the parts in 3 3 3 3 4. All these permutations produce the 45 rhythms in table 6. Including 1 as one of the allowed parts produces one additional partition: 1 3 4 4 4. Including both 1 and 5 as allowed parts produces a total of 11 partitions. They are shown in table 10.

There are four programs for creating partitions. `part.c` creates all partitions of n into any number of parts. `parta.c` creates all partitions of n into any number of parts using a set of allowed parts. `partm.c` creates all partitions of n into m parts. `partam.c` creates all partitions of n into m parts using a set of allowed parts. The partitions in table 10 were created with `partam.c` using the command line `partam 16 5 1 2 3 4 5`.

If you generate all permutations of each of the 11 rhythms in table 10 you get the set of 365 rhythms that we mentioned previously in the section on restricted rhythms. You do not have to use all the permutations. If you restrict the permutations to those that are unique with respect to cyclic shifts you get the rhythm necklaces. You can also fix some of the parts and permute the

Intervals
1 1 4 5 5
1 2 3 5 5
1 2 4 4 5
1 3 3 4 5
1 3 4 4 4
2 2 2 5 5
2 2 3 4 5
2 2 4 4 4
2 3 3 3 5
2 3 3 4 4
3 3 3 3 4

Table 10: Partitions of 16 into 5 parts of size 1, 2, 3, 4, or 5.

others around them. For example with the partition 2 2 3 4 5 you can construct 6 rhythms by setting the first and third interval to 2 and then permuting the 3, 4, and 5 among the remaining intervals in 6 different ways. The program `permi.c` will generate all the permutations of a set of intervals for you.

The idea here is that if you start with partitions then you can create arbitrary rules for defining rhythm sets. Before moving on we should mention that another way to generate sets of rhythms with specific rules is to use what are called finite automata. We will not go into that here, but the interested reader can see our book on finite automata listed in the references section.

DE BRUIJN RHYTHMS

Now we're going to look at a very rich set of complex rhythms which we call de Bruijn rhythms. They are named after *Nicolaas Govert de Bruijn* (1918-2012), a Dutch mathematician. He did a lot of work in the field of combinatorics where he discovered a type of binary sequence called, appropriately enough, a de Bruijn sequence. We are going to use these sequences as rhythms. We'll start with a short description of the theory behind the sequences which is interesting in itself.

The number of binary sequences of length n is 2^n, since there are two choices for each term in the sequence, and there are n choices to be made. For $n = 3$ there are 8 possible sequences. They are: 000, 001, 010, 011, 100, 101, 110, 111. The question is, can you construct a single binary sequence of length 8 that has all 8 of these sequences of length 3 as subsequences? It turns out that you can, it is called a de Bruijn sequence. The sequence is 11101000. Starting at the first term, and taking the next two terms gives you 111. Starting at the second term, and taking the next two terms gives you 110. Continuing on like this gives you 101, 010, 100, and 000. At this point, you treat the string as though it wraps around to the beginning, which then gives you the last two substrings 001 and 011.

What we just looked at is a de Bruijn sequence of order

Figure 1: Nicolaas Govert de Bruijn (1918-2012). Photo credit: Mathematisches Forschungsinstitut Oberwolfach GmbH

3. It turns out there are de Bruijn sequences of all orders. Let $B(n)$ represent a de Bruijn sequence of order n. If you treat $B(n)$ as a circular sequence, where the end wraps around to the beginning, then it will contain all possible binary sequences of length n as subsequences. Since there are 2^n such sequences, $B(n)$ must have length 2^n. For $n > 2$, more than one unique $B(n)$ sequence exists. By unique, we mean that one sequence is not simply a circular shift of another. For $n = 1$, the only sequence is $B(1) = 10$. For $n = 2$, the only sequence is $B(2) = 110$. For $n = 3$, there are two sequences: 11101000 and 11100010. In general, there are $2^{2^{n-1}-n}$ unique $B(n)$ sequences. Table 11 shows the number of $B(n)$ sequences for $n = 1$ to 7 and the length of each sequence.

n	length	number of sequences
1	2	1
2	4	1
3	8	2
4	16	16
5	32	2048
6	64	67108864
7	128	144115188075855872

Table 11: Number and length of $B(n)$ sequences.

The 16 unique $B(4)$ sequences and the intervals between onsets (1's) are shown in table 12. These sequences produce an interesting set of rhythms. We've taken each sequence and played it simultaneously with all possible shifts of itself. This produces very rich rhythm sets. Some of the more interesting ones are shown below and can be listened to on the book's website. The percussion instruments used are low tom and high tom.

$B(4)$	Intervals
1111010110010000	1 1 1 2 2 1 3 5
1101011110010000	1 2 2 1 1 1 3 5
1111011001010000	1 1 1 2 1 3 2 5
1101111001010000	1 2 1 1 1 3 2 5
1011110011010000	2 1 1 1 3 1 2 5
1111001011010000	1 1 1 3 2 1 2 5
1011001111010000	2 1 3 1 1 1 2 5
1100101111010000	1 3 2 1 1 1 2 5
1011110100110000	2 1 1 1 2 3 1 5
1111010010110000	1 1 1 2 3 2 1 5
1001111010110000	3 1 1 1 2 2 1 5
1010011110110000	2 3 1 1 1 2 1 5
1011010011110000	2 1 2 3 1 1 1 5
1101001011110000	1 2 3 2 1 1 1 5
1001101011110000	3 1 2 2 1 1 1 5
1010011011110000	2 3 1 2 1 1 1 5

Table 12: The 16 unique $B(4)$ sequences.

$B(4)$ sequence 1 and its right shift by 5

■■■■□■□■■□□■□□□□

■□□□□■■■■□■□■■□□

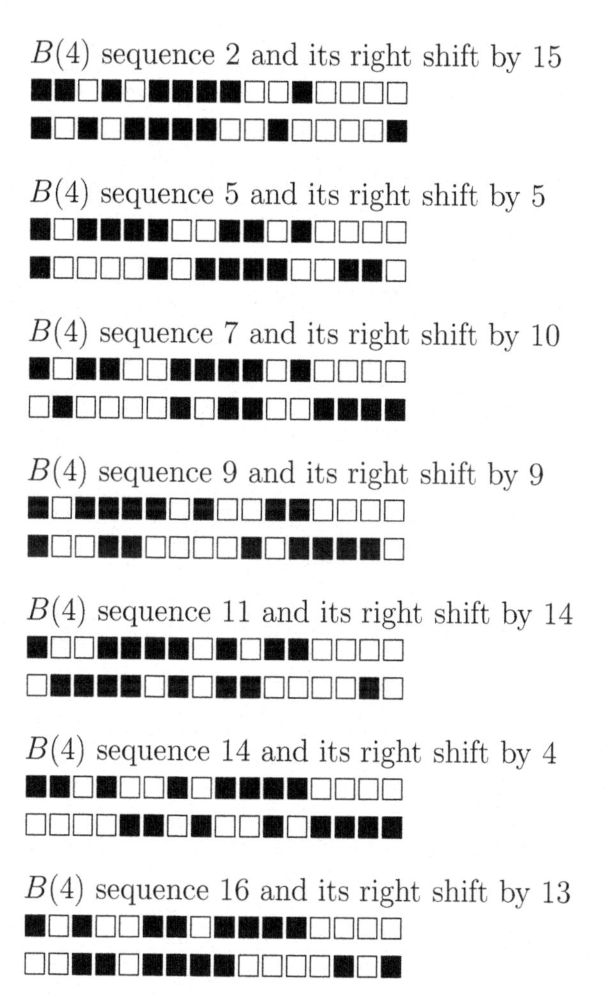

The program `debruijn.c` will produce the largest de Bruijn sequence (largest in the sense of the sequence being a binary number) for any given order. It won't produce all 2048 sequences of order 5, but you can get additional sequences by reversing the one it produces and also by reversing some of its subsequences. As long as you check that all possible binary sequences of length 5 are contained within the sequence then it is a de Bruijn sequence and you have a new rhythm.

CHRISTOFFEL RHYTHMS

Christoffel words are binary sequences that can be used to produce some of the most interesting rhythms in music. They are equivalent (up to cyclic rotation) to what Godfried Toussaint calls Euclidean rhythms. The idea is to distribute a given number of onsets as evenly as possible in a given number of pulses or beats. We will begin by explaining exactly what Christoffel words are and how to construct them. We then give some examples of combining them to produce rhythm sets.

Christoffel words are named after *Elwin Bruno Christoffel* (1829-1900), a German mathematician and physicist. They originated in the study of continued fractions and have since found uses in other areas of mathematics. They are part of the study of the combinatorics of words. In mathematics a word is simply a string of symbols chosen from an alphabet. Christoffel words can be constructed from any binary alphabet (an alphabet of only two symbols) but we will stick to using the numbers 0 and 1 where a 1 indicates a note onset. For an excellent summary of the mathematics of Christoffel words see reference Berstel et al.. We're going to keep the mathematics to the minimum necessary to understand how Christoffel words (C-words) are constructed.

The simplest geometric method for constructing C-

Figure 2: Elwin Bruno Christoffel (1829-1900). Photo credit: wikipedia.org

words uses a grid of equally spaced horizontal and vertical lines like in figure 3. In mathematics this is called a Cartesian grid. Vertical lines are numbered left to right starting with zero. Horizontal lines are numbered bottom to top starting with zero. The point where a vertical and horizontal line intersect is labeled by the respective line numbers which are called the coordinates of the point. Three of these intersection points, $(5,4)$, $(3,1)$, and $(6,2)$ are labeled in the figure.

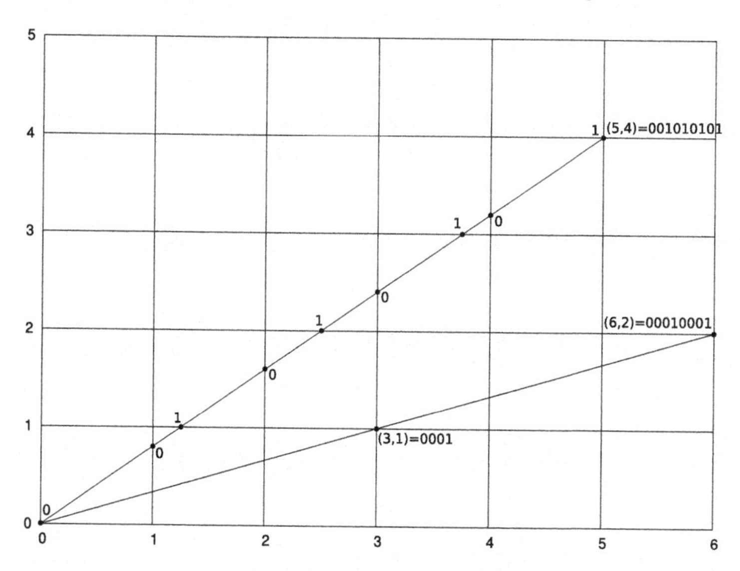

Figure 3: Constructing Christoffel Words.

Strictly speaking, only points whose coordinates are relatively prime have Christoffel words associated with them. By relatively prime, we mean that the two numbers have no common factors. The points $(5,4)$ and

$(3,1)$ both have relatively prime coordinates, but the point $(6,2)$ does not, the coordinates have a common factor of 2. We will stretch the definition of a Christoffel word so that every intersection point has a word associated with it. The implications of this are discussed below.

So let's construct our first Christoffel word. We will construct the word for point $(5,4)$. Start by drawing a line from $(0,0)$ to $(5,4)$ as shown in the figure. As you move along the line from $(0,0)$ to $(5,4)$ it will intersect a series of vertical and horizontal lines. Label each vertical line intersection with a 0 and each horizontal line intersection with a 1. Now if you label the starting point 0 and the ending point 1 then the sequence of 0's and 1's is 001010101 which is the lower Christoffel word for $(5,4)$. If you label the starting point 1 and the ending point 0, then you get 101010100 which is the upper Christoffel word for $(5,4)$. Christoffel words come in two flavors, upper and lower. The upper word always starts with a 1 and ends with a 0, while the opposite is true for the lower word. Between the beginning and the end, the string is always a palindrome, meaning it reads the same in both directions. It also means that the upper word is a reverse of the lower word.

Another way to construct the Christoffel word, which shows where the upper and lower terminology comes from, is to move from point $(0,0)$ to $(5,4)$ using only horizontal and vertical steps along grid lines. To construct the lower word, you take steps that keep you as close as possible below the line from $(0, 0)$ to $(5, 4)$ without crossing it. To construct the upper word, you take steps that keep you as close as possible above the line without crossing it. If horizontal steps are labeled 0 and vertical steps are labeled 1 then the upper and lower paths give you the upper and lower Christoffel words 001010101 and 101010100 for $(5, 4)$ as in the above construction.

A detailed example of this form of construction for both the upper and lower words of point $(11, 5)$ is shown in the figure below.

You can construct a Christoffel word for any point on the grid and each word will be unique. No two points have the same word. However, the word for some points will be repetitions of the word for another point. These repetitions occur when the coordinates of a point are not relatively prime. As we mentioned above, the word in this case is not, strictly speaking, a Christoffel word but we will treat it as one. For example the lower word for $(6, 2)$ is 00010001, which is two repetitions of the lower word $(3, 1) = 0001$. Notice in this case that $(6, 2) = (2 \cdot 3, 2 \cdot 1) = 2 \cdot (3, 1)$ so $(6, 2)$ is a multiple of 2 times $(3, 1)$.

If a generic point (x, y) is a multiple n of another point, so that $(x, y) = (n \cdot a, n \cdot b) = n \cdot (a, b)$ then the word for (x, y) will be composed of n copies of the word for

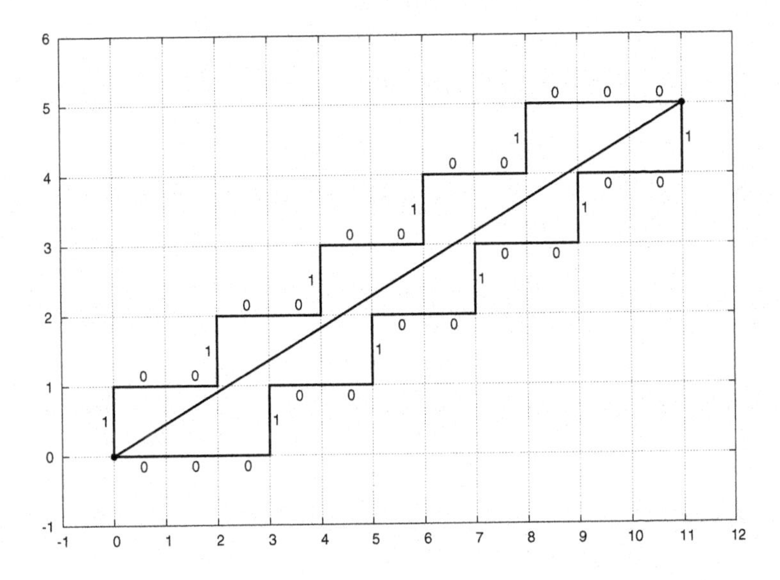

Figure 4: Upper and lower Christoffel words for $(11, 5)$.

(a, b). If you want to avoid words with repetitions then you must pick a point (x, y) where the numbers x and y are relatively prime, i.e. they have no common factors, so that you cannot write them as $x = n \cdot a$ and $y = n \cdot b$ for some integers a, b, and n.

Another way to characterize Christoffel words is in terms of the slope of a line. A line from $(0, 0)$ to point (x, y) will have a slope of y/x. The slope is essentially the steepness of the line. It is the ratio of how far you move in the vertical direction for a given amount of movement in the horizontal direction while traveling along the line. The larger the slope the steeper the line. If point A is a multiple of point B then the line from $(0, 0)$ to A has the same slope as the line from $(0, 0)$ to B so they are really the same lines and the Christoffel word for A is a repeat of the word for B. You can see this for $A = (6, 2)$ and $B = (3, 1)$ in figure 3.

We will usually refer to Christoffel words in terms of the slope, y/x of a line starting from the origin. The lower C-word will be denoted as $L(x, y)$ and the upper C-word as $U(x, y)$. The assumption will always be that both y and x are integers so the slope is a rational number. If the slope were an irrational number such as $\sqrt{2}$ then the line would never intersect another lattice point and the Christoffel word would be infinitely long. It would become what the mathematicians call a *Sturmian sequence*. You can always get as close as you want to an irrational slope by using a rational approximation. We will show how to do this in some of the following examples.

One more way to construct Christoffel words is by using the trajectory of a billiard ball on a square billiard table. You launch a billiard ball from the lower left corner of the table in the direction of a line with slope y/x. The ball will bounce off the sides of the table and eventually end up in one of the corners after $x + y - 2$ bounces. An example is shown in figure 5 for $y/x = 4/5$.

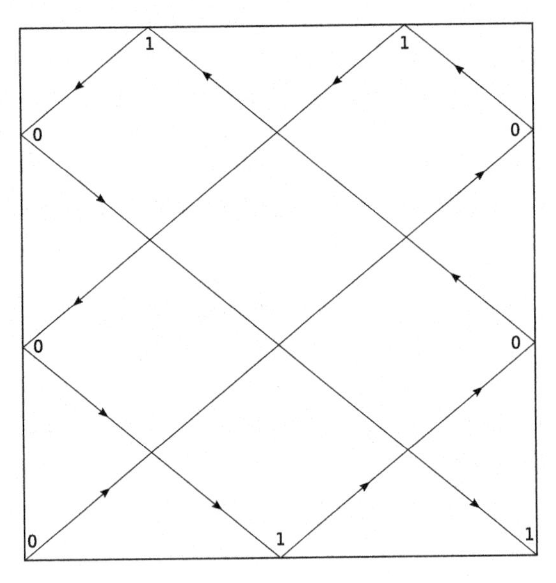

Figure 5: Billiard ball trajectory with initial slope $4/5$.

If you label the start with a 0, a bounce off a vertical

side with a 0, a bounce off a horizontal side with a 1, and the end with a 1, then the trajectory defines the lower Christoffel word for slope y/x. If the start is labeled 1 and the end is labeled 0, then you get the upper Christoffel word. This method can obviously be generalized to trajectories inside shapes other than squares.

Luckily you don't have to manually construct Christoffel words using any of the methods discussed so far. There's a simple algorithm that will generate the words for you. To create the word for y/x start by creating two lists of numbers: $L_0 = [y, 2y, 3y, \ldots, (x-1)y]$ and $L_1 = [x, 2x, 3x, \ldots, (y-1)x]$. Next combine the two lists into a single sorted list. Examine each element of the sorted list. If it comes from L_0 replace it with a 0 and if it comes from L_1 replace it with a 1. Now if you add a 0 to the beginning and a 1 to the end of the list then you have the lower Christoffel word for y/x. Adding a 1 to the beginning and a 0 to the end gives you the upper Christoffel word.

Actually implementing the algorithm is much simpler than it sounds since the creation of the sorted list and conversion to 0's and 1's can be done in a single step. The algorithm in pseudo code is shown below (note that `a != b` means `a` not equal to `b`).

```
a = y
b = x
```

```
if(upper word) print 1 else print 0
while( a != b )
{
  if( a > b )
  {
    print 1
    b = b + x
  }
  else
  {
    print 0
    a = a + y
  }
}
if(upper word) print 0 else print 1
```

The program `chsequl.c` will calculate the C-word for any slope y/x. Run it as:
`chsequl u y x`
and it will print the upper C-word. If you run it with an additional length parameter like this:
`chsequl u y x len`
then it will repeat the C-word out to that length.

From the above algorithm it is clear that y is the number of 1's and x is the number of 0's in both the upper and lower C-word for y/x and you can get the C-word for x/y simply by exchanging the 0's and 1's in the C-word for y/x, leaving the first and last characters unchanged. Also the length of the C-word for y/x will always equal $y + x$. Another thing to keep in mind is that if you remove the first and last characters of any C-word then what you are left with is a palindrome, which is a word that reads the same both forward and backward.

The 1's in a Christoffel word correspond to the note onsets in a rhythm. You can create a C-word rhythm by setting x equal to the number of note onsets that you want, and y equal to $n - x$ where n is the total pulse length of the rhythm. The following tables give upper Christoffel words for different numbers of note onsets and rhythm pulse lengths. These can be directly converted into rhythms. By using upper Christoffel words, the first pulse in the rhythm will always be a note onset.

Table 13 shows the upper C-words for 2 note onsets and pulse lengths from 3 to 16. Note that for even i the number of 0's following each 1 is $i/2$, and for odd i the number of 0's following the first 1 is $(i-1)/2$, and the number of 0's following the second 1 is $(i+1)/2$. For even i, all the intervals are the same, and the rhythm is rather boring. For odd i, there are two intervals that alternate.

These 2 onset C-words can be used to tile a rhythm with a larger number of pulse lengths. Table 14 shows an example where the number of pulses is fixed at 16 and the C-words are repeated to fill the rhythm.

i	$U(i, 2)$	Intervals
1	110	1 2
2	1010	2 2
3	10100	2 3
4	100100	3 3
5	1001000	3 4
6	10001000	4 4
7	100010000	4 5
8	1000010000	5 5
9	10000100000	5 6
10	100000100000	6 6
11	1000001000000	6 7
12	10000001000000	7 7
13	100000010000000	7 8
14	1000000010000000	8 8

Table 13: Upper Christoffel words with 2 onsets.

With tiling, you can get a new interval length when the length of the C-word does not evenly divide the length of the rhythm.

Table 15 shows the upper C-words for 3 note onsets. Note that each word contains at most 2 unique intervals. Words where i is divisible by 3 will only have 1 unique interval. Table 16 shows the effect of using these words to tile a rhythm with 16 pulses.

The following tables show the upper C-words for 4, 5, and 6 onsets.

i	$U(i,2)$	Intervals
1	1101101101101101	1 2 1 2 1 2 1 2 1 2 1
2	1010101010101010	2 2 2 2 2 2 2 2
3	1010010100101001	2 3 2 3 2 3 1
4	1001001001001001	3 3 3 3 3 1
5	1001000100100010	3 4 3 4 2
6	1000100010001000	4 4 4 4
7	1000100001000100	4 5 4 3
8	1000010000100001	5 5 5 1
9	1000010000010000	5 6 5
10	1000001000001000	6 6 4
11	1000001000000100	6 7 3
12	1000000100000010	7 7 2
13	1000000100000001	7 8 1
14	1000000010000000	8 8

Table 14: Upper Christoffel words with 2 onsets tiled to 16 pulses.

i	$U(i,3)$	Intervals
1	1110	1 1 2
2	11010	1 2 2
3	101010	2 2 2
4	1010100	2 2 3
5	10100100	2 3 3
6	100100100	3 3 3
7	1001001000	3 3 4
8	10010001000	3 4 4
9	100010001000	4 4 4
10	1000100010000	4 4 5
11	10001000010000	4 5 5
12	100001000010000	5 5 5
13	1000010000100000	5 5 6

Table 15: Upper Christoffel words with 3 onsets.

i	$U(i,3)$	Intervals
1	1110111011101110	1 1 2 1 1 2 1 1 2 1 1 2
2	1101011010110101	1 2 2 1 2 2 1 2 2 1
3	1010101010101010	2 2 2 2 2 2 2 2
4	1010100101010010	2 2 3 2 2 3 2
5	1010010010100100	2 3 3 2 3 3
6	1001001001001001	3 3 3 3 3 1
7	1001001000100100	3 3 4 3 3
8	1001000100010010	3 4 4 3 2
9	1000100010001000	4 4 4 4
10	1000100010000100	4 4 5 3
11	1000100001000010	4 5 5 2
12	1000010000100001	5 5 5 1
13	1000010000100000	5 5 6

Table 16: Upper Christoffel words with 3 onsets tiled to 16 pulses.

i	$U(i,4)$	Intervals
1	11110	1 1 1 2
2	110110	1 2 1 2
3	1101010	1 2 2 2
4	10101010	2 2 2 2
5	101010100	2 2 2 3
6	1010010100	2 3 2 3
7	10100100100	2 3 3 3
8	100100100100	3 3 3 3
9	1001001001000	3 3 3 4
10	10010001001000	3 4 3 4
11	100100010001000	3 4 4 4
12	1000100010001000	4 4 4 4

Table 17: Upper Christoffel words with 4 onsets.

i	$U(i,4)$	Intervals
1	1111011110111101	1 1 1 2 1 1 1 2 1 1 1 2 1
2	1101101101101101	1 2 1 2 1 2 1 2 1 2 1
3	1101010110101011	1 2 2 2 1 2 2 2 1 1
4	1010101010101010	2 2 2 2 2 2 2 2
5	1010101001010101	2 2 2 3 2 2 2 1
6	1010010100101001	2 3 2 3 2 3 1
7	1010010010010100	2 3 3 3 2 3
8	1001001001001001	3 3 3 3 3 1
9	1001001001000100	3 3 3 4 3
10	1001000100100010	3 4 3 4 2
11	1001000100010001	3 4 4 4 1
12	1000100010001000	4 4 4 4

Table 18: Upper Christoffel words with 4 onsets tiled to 16 pulses.

i	$U(i,5)$	Intervals
1	111110	1 1 1 1 2
2	1110110	1 1 2 1 2
3	11011010	1 2 1 2 2
4	110101010	1 2 2 2 2
5	1010101010	2 2 2 2 2
6	10101010100	2 2 2 2 3
7	101010010100	2 2 3 2 3
8	1010010100100	2 3 2 3 3
9	10100100100100	2 3 3 3 3
10	100100100100100	3 3 3 3 3
11	1001001001001000	3 3 3 3 4

Table 19: Upper Christoffel words with 5 onsets.

i	$U(i,5)$	Intervals
1	1111101111101111	1 1 1 1 2 1 1 1 1 2 1 1 1 1
2	1110110111011011	1 1 2 1 2 1 1 2 1 2 1 1
3	1101101011011010	1 2 1 2 2 1 2 1 2 2
4	1101010101101010	1 2 2 2 2 1 2 2 2
5	1010101010101010	2 2 2 2 2 2 2 2
6	1010101010010101	2 2 2 2 3 2 2 1
7	1010100101001010	2 2 3 2 3 2 2
8	1010010100100101	2 3 2 3 3 2 1
9	1010010010010010	2 3 3 3 3 2
10	1001001001001001	3 3 3 3 3 1
11	1001001001001000	3 3 3 3 4

Table 20: Upper Christoffel words with 5 onsets tiled to 16 pulses.

i	$U(i,6)$	Intervals
1	1111110	1 1 1 1 1 2
2	11101110	1 1 2 1 1 2
3	110110110	1 2 1 2 1 2
4	1101011010	1 2 2 1 2 2
5	11010101010	1 2 2 2 2 2
6	101010101010	2 2 2 2 2 2
7	1010101010100	2 2 2 2 2 3
8	10101001010100	2 2 3 2 2 3
9	101001010010100	2 3 2 3 2 3
10	1010010010100100	2 3 3 2 3 3

Table 21: Upper Christoffel words with 6 onsets.

i	$U(i,6)$	Intervals
1	1111110111111011	1 1 1 1 1 2 1 1 1 1 1 2 1 1
2	1110111011101110	1 1 2 1 1 2 1 1 2 1 1 2
3	1101101101101101	1 2 1 2 1 2 1 2 1 2 1
4	1101011010110101	1 2 2 1 2 2 1 2 2 1
5	1101010101011010	1 2 2 2 2 2 1 2 2
6	1010101010101010	2 2 2 2 2 2 2 2
7	1010101010100101	2 2 2 2 2 3 2 1
8	1010100101010010	2 2 3 2 2 3 2
9	1010010100101001	2 3 2 3 2 3 1
10	1010010010100100	2 3 3 2 3 3

Table 22: Upper Christoffel words with 6 onsets tiled to 16 pulses.

Word Combinations for Rhythm Sets

The 16 pulse rhythms in tables 14, 16, 18, 20, and 22, as well as additional 16 pulse length upper Christoffel words with up to 15 onsets, can be combined to produce some interesting sounding rhythm sets. Using sets of three, a few combinations that we found interesting are listed below. The first, second, and third rhythms of each set uses high tom, electric snare, and cowbell, respectively. They can be listened to on the book's website.

Christoffel 3-word combination rhythm set 1

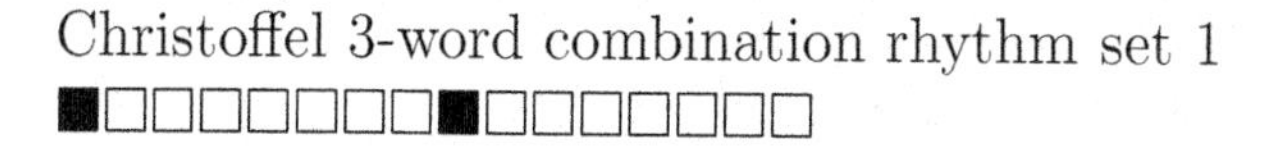

■□□□■□□□■□□□■□□□

■□■□■□■□□■□■□■□□

Christoffel 3-word combination rhythm set 2

■□□□□■□□□□■□□□□□

■□■□□■□□■□■□□■□□

■■□■□■□■■□■□■□■□

Christoffel 3-word combination rhythm set 3

■□□□□□□□■□□□□□□□

■□□■□□■□□■□□■□□□

■■□■■□■□■■□■■□■□

Christoffel 3-word combination rhythm set 4

■□□□□□□□■□□□□□□□

■□□□■□□□■□□□■□□□

■■■□■■□■■□■■□■■□

Christoffel 3-word combination rhythm set 5

■□□□□□□□■□□□□□□□

■□□■□□■□□■□□■□□□

■■■□■■■□■■■□■■■□

Christoffel 3-word combination rhythm set 6

■■■■■□■■■■□■■■■□

■□□□□□□□■□□□□□□□

■□□■□□■□□■□□■□□□

Christoffel 3-word combination rhythm set 7

■■■■■■■□■■■■■■■□

■□□□□■□□□□■□□□□□

■□□■□□■□□■□□■□□□

Christoffel 3-word combination rhythm set 8

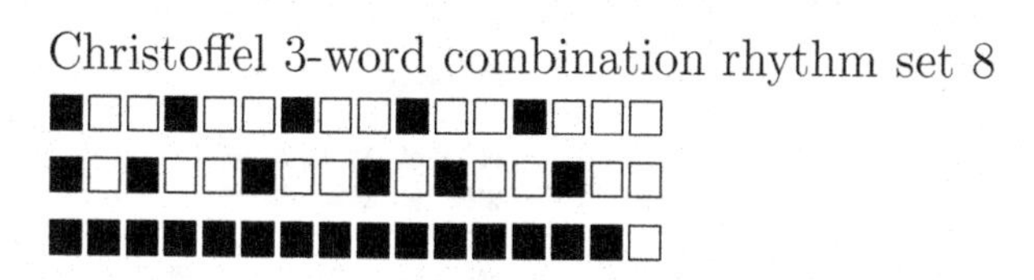

Euclidean Rhythms

It was discovered by Godfried Toussaint that many popular rhythms can be constructed by a process that mimics the Euclidean algorithm for finding the greatest common divisor of two numbers. He calls these Euclidean rhythms and lists many of them in his book. All Euclidean rhythms are equivalent to Christoffel rhythms, either directly or by a cyclic permutation of the intervals. For example Toussaint classifies the Cuban Tresillo rhythm, 10010010, as a Euclidean rhythm. The corresponding Christoffel rhythm is $U(5,3) = 10100100$ which under a cyclic shift becomes the Tresillo. Likewise, he lists the Cuban Cinquillo rhythm as 10110110. The corresponding Christoffel rhythm is $U(3,5) = 11011010$, a cyclic shift of the Cinquillo. Below is a list of some other Euclidean rhythms and their corresponding Christoffel rhythms.

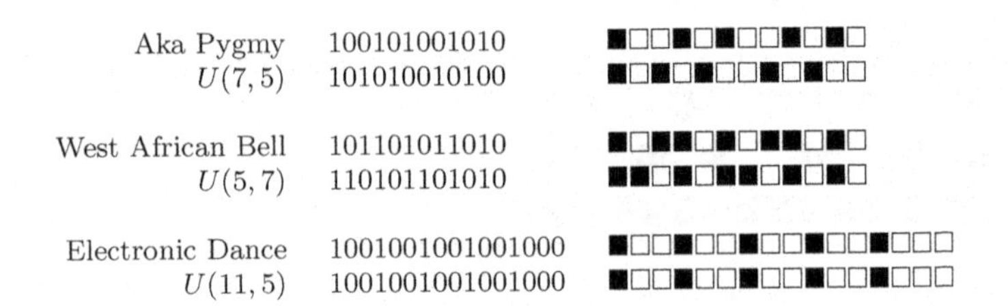

Aka Pygmy	100101001010	■□□■□■□□■□■□
$U(7,5)$	101010010100	■□■□■□□■□■□□
West African Bell	101101011010	■□■■□■□■■□■□
$U(5,7)$	110101101010	■■□■□■■□■□■□
Electronic Dance	1001001001001000	■□□■□□■□□■□□■□□□
$U(11,5)$	1001001001001000	■□□■□□■□□■□□■□□□

Generalized Christoffel Words

Many more rhythms can be created by generalizing how Christoffel words are constructed. Recall how the Christoffel word for the point (x, y) can be constructed as a path from $(0, 0)$ to (x, y) using horizontal and vertical steps. Figure 4 shows the example for point $(11, 5)$. The paths for the upper and lower C-words were not allowed to touch the diagonal line from $(0, 0)$ to $(11, 5)$ except at the end points. If we relax this restriction and allow the path to cross the diagonal line, then many more rhythms can be created.

To show how this is done, let's look at some examples of famous rhythms. Table 23 is a list of famous 16 pulse rhythms. Probably the most famous is the Son Clave. We use this as our first example. Let 1 represent a vertical step, and 0 a horizontal step, then the Son Clave, $U(11, 5)$, and $L(11, 5)$ trace out the paths shown in figure 6. The Son Clave path is almost identical with the $U(11, 5)$ path, except for two points where it crosses the diagonal line. It crosses once to point $(7, 3)$ on the

$L(11,5)$ path, and then it crosses back over again to continue on the $U(11,5)$ path.

This is what we mean by generalizing the Christoffel word construction. The path from $(0,0)$ to (x,y) can cross the diagonal any number of times, as long as it stays within the region bounded by the $U(x,y)$ and $L(x,y)$ paths. All the rhythms in table 23 are generalized C-words, except for the last, which strays outside the region by one point. You could conceivably generalize C-words further by allowing the path to move no more than one step outside the region.

You don't have to go through the process of tracing paths on a grid to get a generalized C-word. You can get one by taking a substring of $U(x,y)$ or $L(x,y)$ and reversing it. Another way is to do a cyclic shift of $U(x,y)$ or $L(x,y)$. Combining a C-word with a shifted version of itself can produce some interesting effects. You can listen to the rhythms in table 23 on the book's website.

Name	Rhythm	Intervals
Bossa Nova	1001001000100100	3 3 4 3 3
Gahu	1001001000100010	3 3 4 4 2
Rumba Clave	1001000100101000	3 4 3 2 4
Shiko	1000101000101000	4 2 4 2 4
Son Clave	1001001000101000	3 3 4 2 4
Soukous	1001001000110000	3 3 4 1 5

Table 23: Famous 16 pulse rhythms.

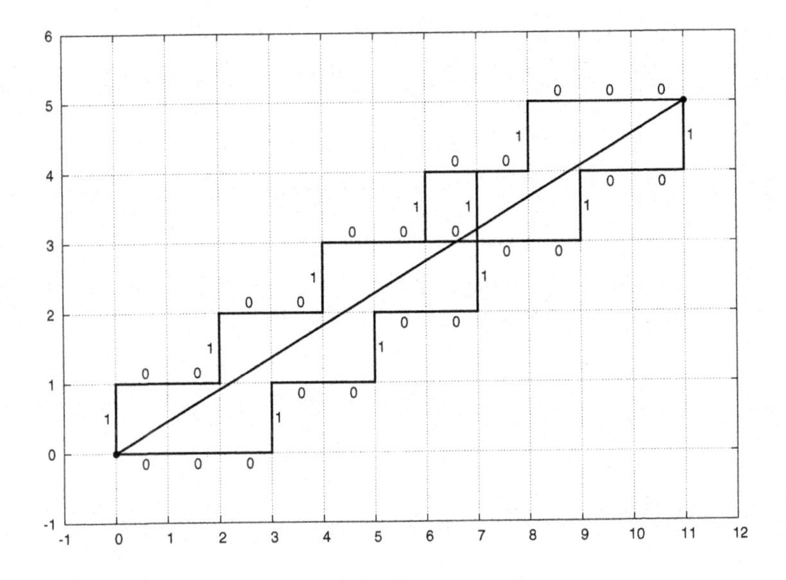

Figure 6: Son Clave and $U(11, 5)$

Christoffel Word Sequences

A Christoffel word is only defined for rational numbers, since it is a finite sequence of 1's and 0's. The Christoffel word for an irrational number would be infinitely long, which would make it what is called a Sturmian sequence. It is possible however to look at Christoffel words for rational approximations of an irrational number. These will form sequences of Christoffel words that are related and can be combined to produce interesting rhythms.

Take the square root of 2 as an example, its value to 20 decimal places is $\sqrt{2} = 1.41421356237309504880\ldots$. A table of $\sqrt{2}$ along with some increasingly accurate rational approximations is shown below.

$\sqrt{2}$	1.414213562373095...
$\frac{7}{5}$	1.4
$\frac{17}{12}$	1.41666666666666
$\frac{41}{29}$	1.41379310344827
$\frac{99}{70}$	1.41428571428571
$\frac{239}{169}$	1.41420118343195
$\frac{577}{408}$	1.41421568627451

These rational approximations come from the continued fraction for $\sqrt{2}$. A continued fraction is like a regular fraction except that the denominator also contains a fraction, and that fraction's denominator also contains a fraction, and so on. A simple continued fraction looks like the following

$$a_0 + \cfrac{1}{a_1 + \cfrac{1}{a_2 + \cfrac{1}{a_3 + \cfrac{1}{a_4 + \cdots}}}}$$

This is usually abbreviated by just listing the numbers like so: $[a_0, a_1, a_2, a_3 \ldots]$. Both rational and irrational numbers have continued fractions. For a rational number, the continued fraction ends at some point, while for an irrational number, it continues on forever. The square root of an integer that is not a perfect square is irrational and its continued fraction goes on forever, but at some point it begins to repeat. The continued fraction for $\sqrt{n}$ will look like $[a_0, \overline{a_1, a_2, \ldots, a_k}]$ where the bar over the terms following a_0 indicate that they repeat forever. So if $k = 2$ the continued fraction would be $[a_0, a_1, a_2, a_1, a_2, \ldots]$. Two interesting facts about these square root continued fractions is that the a_k term will always be twice the a_0 term, and if you remove the a_k term, the remaining terms in the repeating part form a palindrome.

Let's look at the continued fraction for a rational number like 7/5. Start by writing

$$\frac{7}{5} = 1 + \frac{2}{5} = 1 + \frac{1}{\frac{5}{2}} = 1 + \frac{1}{2 + \frac{1}{2}}$$

Putting this in list form, we have $7/5 = [1, 2, 2]$.

Doing the same basic thing for $\sqrt{2}$ produces

$$\sqrt{2} = 1 + \cfrac{1}{2 + \cfrac{1}{2 + \cfrac{1}{2 + \cfrac{1}{2 + \cdots}}}}$$

Here the continued fraction goes on forever with the 2 repeating, which we write as $\sqrt{2} = [1, \overline{2}]$. The bar over the 2 signifies that the 2 repeats forever. Note that the 7/5 example that we worked out contains the first 3 terms of the continued fraction for $\sqrt{2}$ and that 7/5 is one of the rational approximations that we listed above for $\sqrt{2}$. In general to get a rational approximation for a square root, you use a finite number of the beginning terms of its continued fraction. So the next rational approximation for $\sqrt{2}$ is $17/12 = [1, 2, 2, 2]$ and so on.

These increasingly accurate rational approximations are called convergents. The C-words for the first 4 convergents of $\sqrt{2}$ are listed in table 24.

y/x	$U(x,y)$
1/1	10
3/2	11010
7/5	110101101010
17/12	11010110101101010110101101010

Table 24: C-words for convergents of $\sqrt{2}$.

Let's create rhythm sets with the C-word of the fourth convergent, 17/12, which has a length of 29. We'll make four sets with two percussion instruments: high and low bongos. The high bongo will play the original rhythm, and the low bongo will play a copy that is a cyclic shift to the left by one, two, three and four pulses, resulting in the following four rhythm sets that you can listen to on the book's website. You'll notice that they all sound significantly different from each other, in spite of them being successively different only by a shift of one.

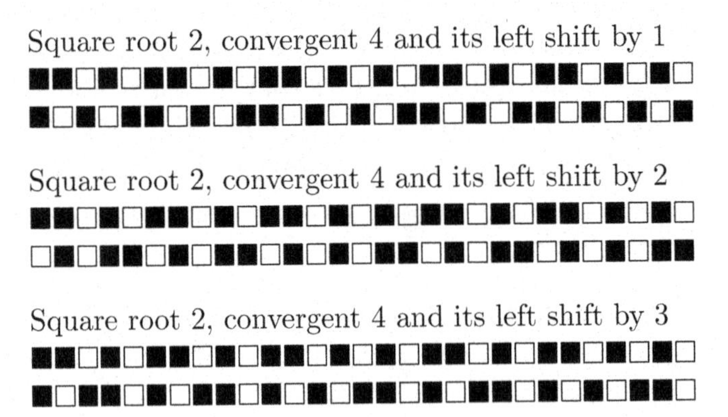

Square root 2, convergent 4 and its left shift by 4

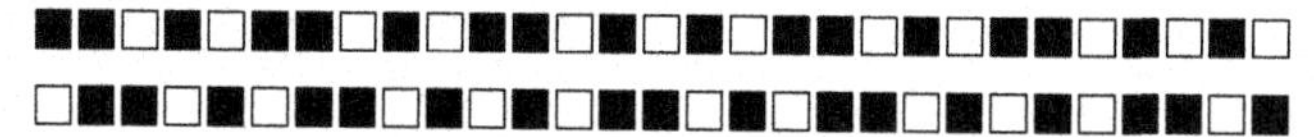

Square Root of 3

Appendix B lists continued fractions for square roots of integers less than 100. You can also use the program `cfsqrt.c` to calculate the continued fraction for the square root of any integer. Run it as

```
cfsqrt n
```

where `n` is the integer whose square root you want the continued fraction for. Running the program for `n`$= 3$ prints out the three numbers 1 1 2 so the continued fraction is $\sqrt{3} = [1, \overline{1, 2}]$. Note that for any square root the numbers following the first number will be the repeating part of the continued fraction. To find the convergents you can run the program `cfcv.c` with however many terms of the continued fraction you want to use. The more terms, the more accurate the convergent. For $\sqrt{3}$ running

```
cfcv 1 1 2
```

will output `5 3` which is the convergent $5/3$ for the first 3 terms of the continued fraction. To add two more terms run

```
cfcv 1 1 2 1 2
```

which outputs `19 11` which is the convergent $19/11$ for

the first 5 terms of the continued fraction. The C-words for the first 5 convergents of $\sqrt{3}$ are listed in table 25.

y/x	$U(x,y)$
1/1	10
2/1	110
5/3	11011010
7/4	11011011010
19/11	110110110101101101011011010

Table 25: C-words for convergents of $\sqrt{3}$.

As we did with $\sqrt{2}$, let's take the $\sqrt{3}$ convergent, 19/11, and create rhythm sets with its C-word, which has a length of 30. Again, we'll make four sets with two instruments: high and low bongos, with each set having a cyclic shift to the left by one, two, three and four pulses, which you can listen to on the book's website.

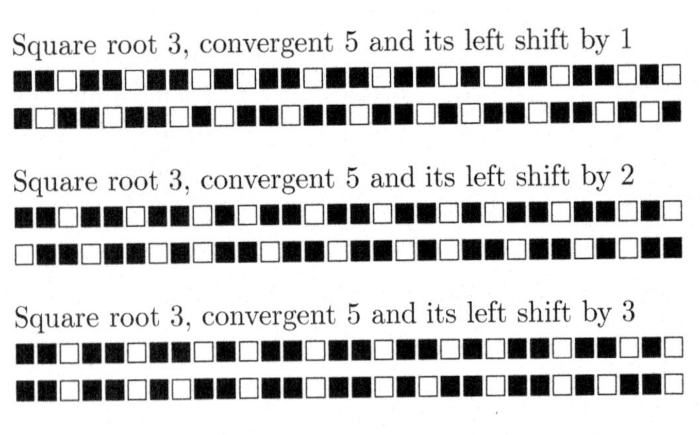

Square root 3, convergent 5 and its left shift by 4

Square Root of 5

We will look at a couple more examples using square root convergents before moving on to more general irrational numbers. Next, let's look at $\sqrt{5}$. The continued fraction is $\sqrt{5} = [2, \overline{4}]$ and the C-words for the first 3 convergents of $\sqrt{5}$ are listed in table 26.

y/x	$U(x, y)$
2/1	110
9/4	1110110110110
38/17	1110110110110111011011011011101101101101110110110110110

Table 26: C-words for convergents of $\sqrt{5}$.

With the same procedure as previous, let's take the $\sqrt{5}$ convergent, 38/17, and create rhythm sets with its C-word, which has a length of 55. Again, we use high and low bongos, with each set having a cyclic shift to the left by one, two, three and four pulses, which you can listen to on the book's website.

Square root 5, convergent 3 and its left shift by 1

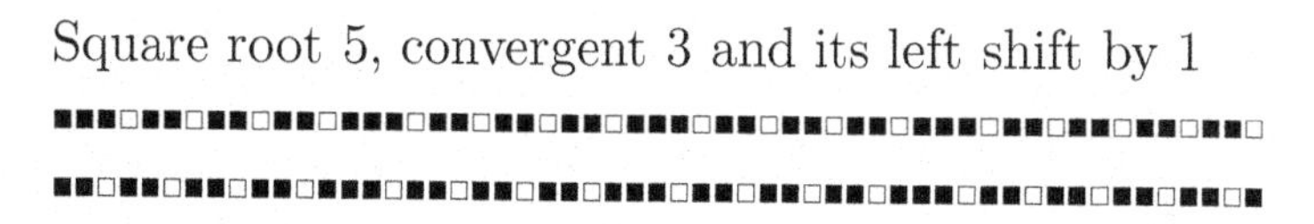

Square root 5, convergent 3 and its left shift by 2

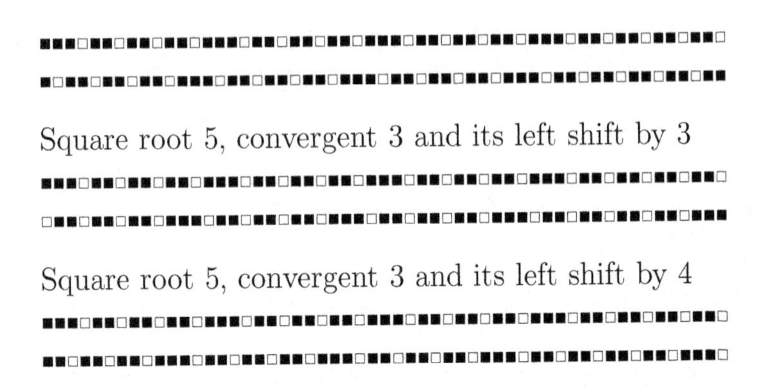

Square root 5, convergent 3 and its left shift by 3

Square root 5, convergent 3 and its left shift by 4

Square Root of 17

As one more example, let's look at $\sqrt{17}$. The continued fraction is $\sqrt{17} = [4, \overline{8}]$ and the C-word for the first 3 convergents of $\sqrt{17}$ are in table 27.

y/x	$U(x, y)$
4/1	11110
33/8	11111011110111101111011110111101111011110
268/65	large string of length 333

Table 27: C-words for convergents of $\sqrt{17}$.

Repeating as previous, let's take the $\sqrt{17}$ convergent, 33/8, and create rhythm sets with its C-word, which has a length of 41. Again, we use high and low bongos, with each set having a cyclic shift to the left by one, two, three and four pulses, which you can listen to on the book's website.

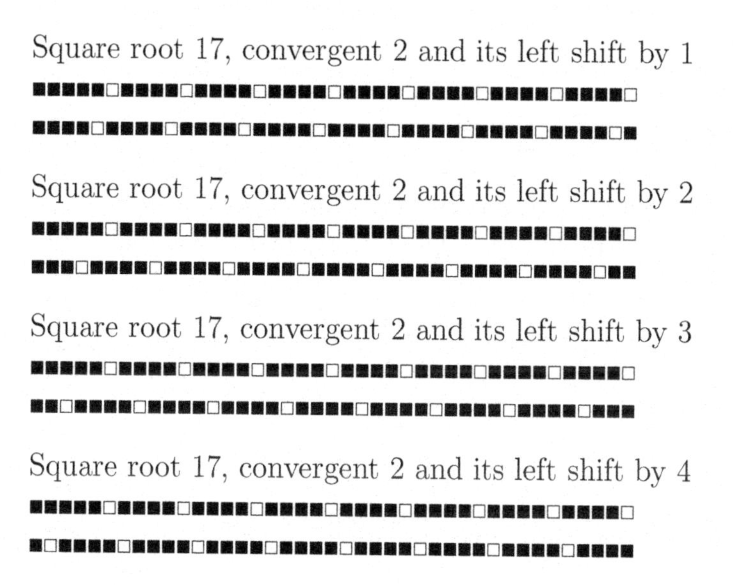

Golden Ratio

Now we look at some more general irrational numbers. The simplest of these are the quadratic irrationals. The name comes from the fact that they are roots of quadratic equations with integer coefficients. These are equations of the form: $ax^2 + bx + c = 0$ where a, b, c are integers. The two solutions (roots) of this equation can be written as: $\frac{-b \pm \sqrt{b^2-4ac}}{2a}$. If the argument inside the square root is positive, meaning $b^2 > 4ac$ then the roots will be quadratic irrationals with periodic continued fractions like the simple square roots[2].

The quadratic irrational with the simplest continued fraction is the golden ratio: $\phi = \frac{1+\sqrt{5}}{2}$ which is the positive root of the quadratic equation $x^2 - x - 1 = 0$. The continued fraction is purely periodic: $\phi = [\overline{1}]$, i.e. all of the infinite number of terms are equal to 1. The C-words for the first few convergents of ϕ are in table 28.

y/x	$U(x, y)$
1/1	10
2/1	110
3/2	11010
5/3	11011010
8/5	1101101011010
13/8	110110101101101011010
21/13	1101101011011010110101101101011010
34/21	1101101011011010110101101101011010110110101101101011010

Table 28: C-words for convergents of ϕ.

Repeating as previous, let's take the ϕ convergent, 21/13, and create rhythm sets with its C-word, which has a length of 34. Again, using high and low bongos, with each set having a cyclic shift to the left by one, two, three and four pulses, you can listen to the rhythm sets on the book's website.

Golden ratio, convergent 7 and its left shift by 1

[2]Simple square roots are, strictly speaking, also quadratic irrationals with $b = 0$, $a = 1$, and $c < 0$, but we are interested here in the more general case.

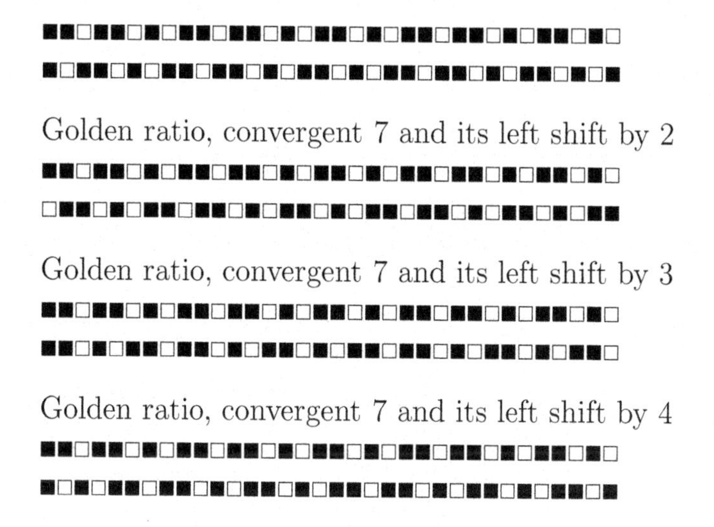

Golden ratio, convergent 7 and its left shift by 2

Golden ratio, convergent 7 and its left shift by 3

Golden ratio, convergent 7 and its left shift by 4

Transcendental Numbers

Another class of irrational numbers are called the transcendental numbers. These numbers are not the roots of any polynomial equation with integer coefficients. Probably the most familiar transcendental number is Pi which is the ratio of a circle's circumference to its diameter and is usually symbolized as π. Its value to 20 decimal places is $\pi = 3.14159265358979323844\ldots$ The continued fraction for π not only goes on forever, but it does not repeat and there is no recognizable pattern to the terms. The first few terms of the continued fraction are
$\pi = [3, 7, 15, 1, 292, 1, 1, 1, 2, 1, 3, 1, 14, \ldots]$.

The C-words for the first few convergents of π are in table 29.

y/x	$U(x, y)$
3/1	1110
22/7	11110111011101110111011101110
333/106	large string of length 439

Table 29: C-words for convergents of π.

Repeating as previous, let's take the π convergent, 22/7, and create rhythm sets with its C-word, which has a length of 29. Again, using high and low bongos, with each set having a cyclic shift to the left by one, two, three and four pulses, you can listen to the rhythm sets on the book's website.

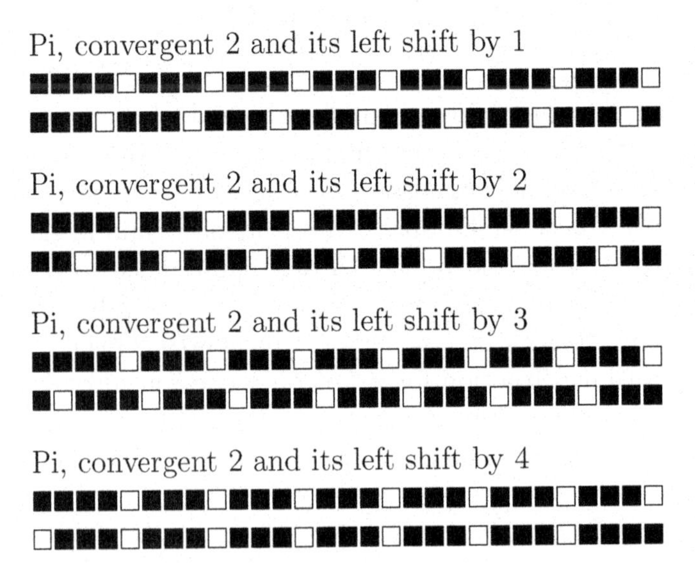

Another transcendental number is e, the base of the natural logarithm. Its value to 20 decimal places is $e = 2.71828182845904523536\ldots$ Its continued fraction is $e = [2, 1, 2, 1, 1, 4, 1, 1, 6, 1, 1, 8, 1, 1, 10, \ldots]$. The pattern of two 1's followed by $4, 6, 8, 10, 12, \ldots$ continues forever. The C-words for the first few convergents of e are in table 30.

y/x	$U(x, y)$
2/1	110
3/1	1110
8/3	11101110110
11/4	111011101110110
19/7	11101110111011011101110110
87/32	large string of length 119

Table 30: C-words for convergents of e.

Repeating as previous, let's take the e convergent, 19/7, and create rhythm sets with its C-word, which has a length of 26. Again, using high and low bongos, with each set having a cyclic shift to the left by one, two, three and four pulses, you can listen to the rhythm sets on the book's website.

e, convergent 5 and its left shift by 1

e, convergent 5 and its left shift by 2

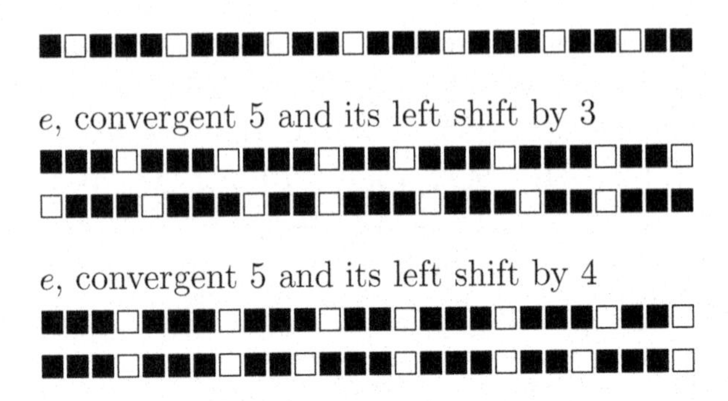

The Euler-Mascheroni constant, or gamma, appears in probability, statistics, physics, number theory, and mathematical analysis. Whether it's an irrational or transcendental number is still not known. Its value to 20 decimal places is
$\gamma = 0.57721566490153286061\ldots$.
Its continued fraction is
$\gamma = [0, 1, 1, 2, 1, 2, 1, 4, 3, 13, 5, 1, 1, 8, \ldots]$.
Like π there is no discernible pattern to the terms and since it is not known for sure if γ is irrational we don't know if the terms go on forever or stop at some point. The odds are good that it is irrational and the terms never stop. The C-words for the first few convergents of γ are in table 31.

Repeating as previous, let's take the γ convergent, 11/19, and create rhythm sets with its C-word, which has a length of 30. Again, using high and low bongos, with each set having a cyclic shift to the left by one, two, three and four pulses, you can listen to the rhythm sets on the book's website.

y/x	$U(x,y)$
0/1	1
1/1	10
1/2	100
3/5	10100100
4/7	10100100100
11/19	101001001010010010010100100100
15/26	10100100101001001001010010010010100100100

Table 31: C-words for convergents of γ.

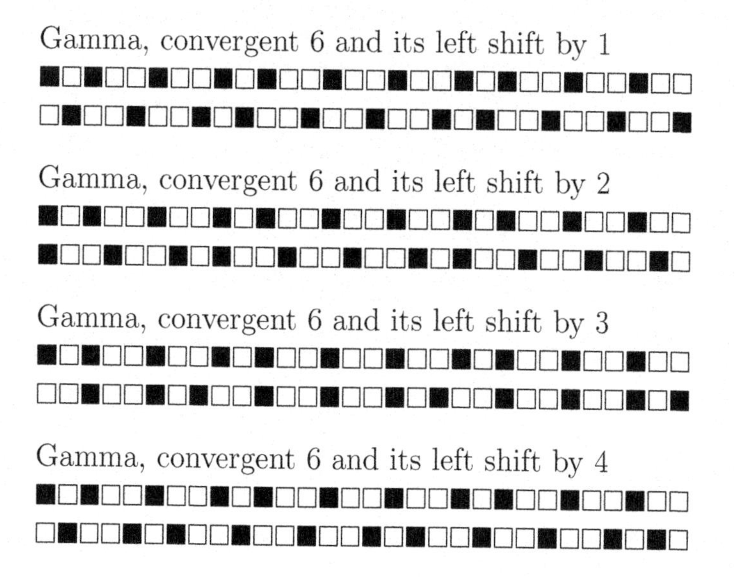

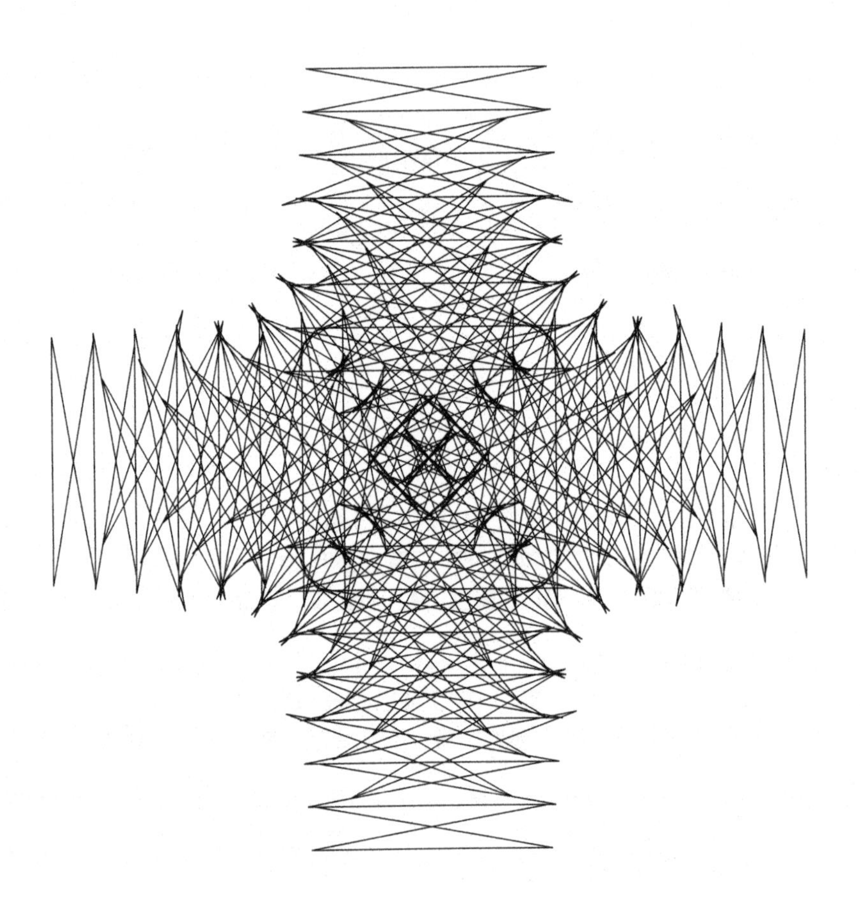

NATURAL RHYTHMS

Here we show how to create rhythm sets from any set of things that are quantifiable. Two examples we look at are the orbital periods of the planets, and the abundance of elements in the universe. Our method is to represent the quantity from which the rhythm set is made as a rational number, get its continued fraction, then use a convergent of the continued fraction, whose sum of numerator and denominator isn't too large, to get the corresponding Christoffel word that is used for the rhythm. In this way, the set of things produces a rhythm set. It's easy to see with examples. First, let's look at the planets.

Rhythms of the Planets

Shown in table 32 are the orbital periods of the nearest six planets to the sun, as fractions of Earth years, as well as the corresponding continued fraction. The continued fraction is gotten with the program `cfrat.c`.

Because the scale of Jupiter and Saturn's periods are much larger, we'll drop those, and only look at the first four planets. Choosing a sufficient number of continued fraction terms to make a convergent that is close to the original period, but whose sum isn't unreasonable in

Planet	Period	Cont frac
Mercury	88/365	[0,4,6,1,3,3]
Venus	225/365	[0,1,1,1,1,1,1,5]
Earth	1/1	[1]
Mars	687/365	[1,1,7,2,21]
Jupiter	4332/365	[11,1,6,1,1,1,1,9]
Saturn	10767/365	[29,2,182]

Table 32: Orbital periods of first six planets as fractions of Earth years, and corresponding continued fraction.

terms of the length of a Christoffel word, we get the convergents shown in table 33. The convergents are gotten with the program `cfcv.c`.

Planet	Terms used	Convergent
Mercury	[0,4,6]	6/25
Venus	[0,1,1,1,1,1,1]	8/13
Earth	[1]	1/1
Mars	[1,1,7]	15/8

Table 33: Continued fraction terms used, and corresponding convergent.

The Christoffel words corresponding to the chosen convergents are shown in table 34. These Christoffel words are upper words calculated with program `chsequl.c`.

The largest Christoffel word in table 34 is that of Mercury, with a length of 31. To make it easier to discern

Planet	Convergent	Christoffel word
Mercury	6/25	1000010000100001000010000100000
Venus	8/13	101001010010010100100
Earth	1/1	10
Mars	15/8	11011011011011011011010

Table 34: Chosen convergent and corresponding upper Christoffel word.

the patterns when listening, we'll take all the Christoffel word lengths out to twice this, or 62. This is easily done by specifying the length as the last parameter to the program `chsequl.c`. The extended Christoffel words are shown below in box notation, which can be listened to as a midi rhythm set on the book's website, with Mercury as cowbell, Venus as high tom, Earth as electric snare, and Mars as bass drum 1.

The Planets

Rhythms of the Elements

The top ten most abundant elements in the universe are shown in table 35. Also shown are the atomic number, the ratio of the atomic number to that of neon, and finally the Christoffel word for that ratio.

Element	At#	At#/neon	C-word
Hydrogen	1	1/10	10000000000
Helium	2	1/5	100000
Oxygen	8	4/5	101010100
Neon	10	1/1	10
Nitrogen	7	7/10	10101001010010100
Carbon	6	3/5	10100100
Silicon	14	7/5	110101101010
Magnesium	12	6/5	11010101010
Iron	26	13/5	111011101101110110
Sulfur	16	8/5	1101101011010

Table 35: Top ten most abundant elements in the universe. Courtesy of Jefferson Lab.

Making a rhythm set from the top seven, by extending the Christoffel words to twice the maximum length ($2 \cdot 17 = 34$) gives the following, with hydrogen as bass drum 1, helium as electric snare, oxygen as low tom, neon as high tom, nitrogen as high bongo, carbon as low bongo, and silicon as cowbell.

The Elements

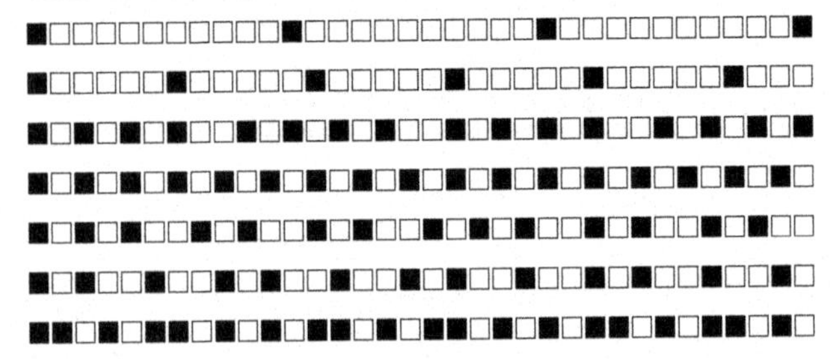

FOLDING RHYTHMS

A large class of binary sequences, and therefore rhythms, can be generated using the idea of paper folding. Put a rectangular strip of paper on a flat surface in front of you with the long dimension going left to right. Now pick up the right end of the paper and fold it over onto the left end. Repeat this process a few times and then unfold the paper. What you will see is a sequence of creases in the paper, some will look like valleys and some will look like ridges. With one fold there will be a single valley in the middle after unfolding. With two folds you will have three creases of type: valley, valley, ridge, from left to right. Three folds gives you 7 creases and in general n folds gives you $2^n - 1$ creases. With a real sheet of paper it is hard to get more than five or six folds but mathematically you can fold as many times as you like and it is easy to predict the resulting sequence of creases.

Let valley creases be symbolized by the number 1 and ridge creases by the number 0 then the sequence of creases after 1, 2, 3, and 4 folds will look like:
1
110
1101100
110110011100100
Looking at this series of sequences you may notice that the next sequence is found by adding a 1 to the end of

the current sequence and then reflecting it about the end point with 1's and 0's interchanged. Take for example the second sequence, 110, and add a 1 to the end to get 1101. Now reverse 110 with 1's and 0's interchanged to get 100. Add this to 1101 to get the next sequence 1101100. For any sequence in the series, this procedure gives you the next sequence. To formalize the process let S_n be the sequence of creases after n folds and let $\overline{S_n^R}$ represent the reverse of the sequence with 1's and 0's interchanged, then the next sequence is $S_{n+1} = S_n\,1\,\overline{S_n^R}$.

The next sequence always has the previous sequence as a prefix, we are just adding new terms onto the end. This means that if you continue folding forever you will produce a unique infinite sequence of creases. In other words, there is a unique S_∞ which is called the regular paper folding sequence. The sequence out to 31 terms is

$$S_\infty = 1101100111001001110110001100 10\ldots$$

It is possible to calculate the individual terms in S_∞ directly. Let c_n be the nth term in the sequence then its value can be calculated directly from the value of n. Any integer n can be expressed in the form

$$n = 2^k(2j+1)$$

The term $2j+1$ is the odd part of n, meaning it is an odd number. When an odd number is divided by 4 it will have a remainder of 1 or 3 which we symbolize as $(2j + 1) \equiv 1 \mod 4$ and $(2j + 1) \equiv 3 \mod 4$. The value of c_n is then given by

$$c_n = \begin{cases} 1 & \text{if } (2j+1) \equiv 1 \mod 4 \\ 0 & \text{if } (2j+1) \equiv 3 \mod 4 \end{cases}$$

The program `pfold.c` will generate the regular paper folding sequence. It takes 3 parameters and is called like this:

```
pfold n m f
```

The first parameter, `n`, is the number of terms you want to calculate. To get the complete sequence for k folds you should set `n`$= 2^k - 1$. The other 2 parameters can be used to generate a variety of different paper folding sequences which we will discuss below. For the regular paper folding sequence set `m`$= 1$ and `f`$= 1$.

The regular paper folding sequence results from folding the paper in the same direction all the time. If you fold the paper in alternating directions, or you use some other series of folds, then you will get a different sequence. If the paper is folded in alternating directions then you will get the following series of sequences after 1, 2, 3, and 4 folds:
1
100

1001110
100111001000110
Like the regular paper folding sequence, you can get the next sequence in the series by adding a term to the end and reflecting the sequence about that term with 1's and 0's interchanged. The difference is that the terms added to the end alternate between 1 and 0. So for example $S_3 = S_2\,1\,\overline{S_2^R}$ while $S_4 = S_3\,0\,\overline{S_3^R}$. In the limit as you do the alternate folding an infinite number of times, you once again get a unique sequence S_∞ called the alternating paper folding sequence.

Instead of using alternating folds, you can use any pattern of folds. Let a_k indicate the direction of the kth fold. It can have two values: 1 or 0. In the regular paper folding sequence, all the a_k are equal to 1. In the alternating paper folding sequence, they alternate between 1 and 0. Given the a_k values, you can calculate the terms in the resulting paper folding sequence using the following formula

$$c_n = \begin{cases} a_k & \text{if } (2j+1) \equiv 1 \mod 4 \\ 1 - a_k & \text{if } (2j+1) \equiv 3 \mod 4 \end{cases}$$

where the k and j values come from factoring n into the form $n = 2^k(2j+1)$.

So all we need to generate general paper folding sequences is a series of folding instructions, the a_k values. The simplest thing to do is to use a periodic sequence of binary values for a_k. These can come from the binary expansion of a given number f which we will call the folding function. To fully specify the folding function, you need not just the value of f but also a size value indicating how many bits are used to represent f. Call the size value m then f will be represented with m bits and can have a value from 0 to $2^m - 1$.

For example if $m = 1$ then f is represented by 1 bit and can have only two values: 0 and 1. The binary representation of these values is also 0 and 1. For the regular paper folding sequence we used $m = 1$ and $f = 1$ so that in binary $f = 1$ and all the a_k values equal 1. The alternating paper folding sequence can be generated with $m = 2$ and $f = 1$ so that in binary $f = 01$ and the a_k values, starting with $k = 0$, are $10101010\ldots$. The value of m is important since it can change how f is represented and thereby change the values of a_k. For example if $m = 4$ and $f = 3$ then f is represented in binary as $f = 0011$, and the a_k values, starting with $k = 0$, are $110011001100\ldots$.

Note that the a_k values, starting with $k = 0$, are found by writing the binary representation of f backwards and repeating it. This means that a_k will equal bit $i = k \mod m$ of f. The program `pfold.c` will do all these calculations for you and generate the resulting paper folding sequence. You call it like this:

```
pfold n m f
```

where `n` is the number of terms to generate, `m` is the size of the binary representation of the folding function and `f` is the folding function number which can range from 0 to $2^m - 1$.

To create some rhythms with paper folding let's run `pfold` as follows:
`pfold 15 4 i`
where `i`, the folding function, will be incremented from 0 to $2^4 - 1 = 15$. The following are the resulting 16 `pfold` binary sequences of length 15 in numeric order of folding function.

output of `pfold`	#1's	`f`
001001100011011	7	0
100011001001110	7	1
011000100111001	7	2
110010001101100	7	3
001101100010011	7	4
100111001000110	7	5
011100100110001	7	6
110110001100100	7	7
001001110011011	8	8
100011011001110	8	9
011000110111001	8	10
110010011101100	8	11
001101110010011	8	12
100111011000110	8	13
011100110110001	8	14
110110011100100	8	15

Sorting in order of the binary sequence gives:

output of `pfold`	#1's	`f`
001001100011011	7	0
001001110011011	8	8
001101100010011	7	4
001101110010011	8	12
011000100111001	7	2
011000110111001	8	10
011100100110001	7	6
011100110110001	8	14
100011001001110	7	1
100011011001110	8	9
100111001000110	7	5
100111011000110	8	13
110010001101100	7	3
110010011101100	8	11
110110001100100	7	7
110110011100100	8	15

Using the 4-bit binary representation of the function number and sorting in the order of a 4-bit binary Gray code, gives:

output of `pfold`	#1's	f
001001100011011	7	0
100011001001110	7	1
110010001101100	7	3
011000100111001	7	2
011100100110001	7	6
110110001100100	7	7
100111001000110	7	5
001101100010011	7	4
001101110010011	8	12
100111011000110	8	13
110110011100100	8	15
011100110110001	8	14
011000110111001	8	10
110010011101100	8	11
100011011001110	8	9
001001110011011	8	8

Note the pattern of 1's, in the Gray code function ordering, forms an X, which you can see better in the box notation below.

boxed output of `pfold`	#1's	f
□□■□□■■□□□■■□■■	7	0
■□□□■■□□■□□■■■□	7	1
■■□□■□□□■■□■■□□	7	3
□■■□□□■□□■■■□□■	7	2
□■■■□□■□□■■□□□■	7	6
■■□■■□□□■■□□■□□	7	7
■□□■■■□□■□□□■■□	7	5
□□■■□■■□□□■□□■■	7	4
□□■■□■■■□□■□□■■	8	12
■□□■■■□■■□□□■■□	8	13
■■□■■□□■■■□□■□□	8	15
□■■■□□■■□■■□□□■	8	14
□■■□□□■■□■■■□□■	8	10
■■□□■□□■■■□■■□□	8	11
■□□□■■□■■□□■■■□	8	9
□□■□□■■■□□■■□■■	8	8

Midi files of some rhythms formed by contiguous pairs and triples from the Gray code function ordering above can be listened to on the book's website.

Similarly, binary sequences of length 31 can be generated by calling `pfold` in this way:

```
pfold 31 5 i
```

where `i`, is now incremented from 0 to $2^5 - 1 = 31$. The following are the resulting 32 `pfold` binary sequences of length 31 sorted by 5-bit binary Gray code function number $(0, 1, 3, 2, 6, \ldots)$:

output of `pfold`	#1's	f
0010011000110110001001110011011	15	0
1000110010011100100011011001110	15	1
1100100011011000110010011101100	15	3
0110001001110010011000110111001	15	2
0111001001100010011100110110001	15	6
1101100011001000110110011100100	15	7
1001110010001100100111011000110	15	5
0011011000100110001101110010011	15	4
0011011100100110001101100010011	15	12
1001110110001100100111001000110	15	13
1101100111001000110110001100100	15	15
0111001101100010011100100110001	15	14
0110001101110010011000100111001	15	10
1100100111011000110010001101100	15	11
1000110110011100100011001001110	15	9
0010011100110110001001100011011	15	8
0010011100110111001001100011011	16	24
1000110110011101100011001001110	16	25
1100100111011001110010001101100	16	27
0110001101110011011000100111001	16	26
0111001101100011011100100110001	16	30
1101100111001001110110001100100	16	31
1001110110001101100111001000110	16	29
0011011100100111001101100010011	16	28
0011011000100111001101110010011	16	20
1001110010001101100111011000110	16	21
1101100011001001110110011100100	16	23
0111001001100011011100110110001	16	22
0110001001110011011000110111001	16	18
1100100011011001110010011101100	16	19
1000110010011101100011011001110	16	17
0010011000110111001001110011011	16	16

Note that again the pattern of 1's, in the Gray code function ordering above forms an X, which you can see better in the box notation below. Midi files of some rhythms formed by contiguous pairs and triples from the Gray code function ordering below can be listened to on the book's website.

boxed output of `pfold`	#1's	f
□□■□□■■□□□■■□■■□□□■□□■■■□□■■□■■	15	0
■□□□■■□□■□□■■■□□■□□□■■□■■□□■■■□	15	1
■■□□■□□□■■□■■□□□■■□□■□□■■■□■■□□	15	3
□■■□□□■□□■■■□□■□□■■□□□■■□■■■□□■	15	2
□■■■□□■□□■■□□□■□□■■■□□■■□■■□□□■	15	6
■■□■■□□□■■□□■□□□■■□■■□□■■■□□■□□	15	7
■□□■■■□□■□□□■■□□■□□■■■□■■□□□■■□	15	5
□□■■□■■□□□■□□■■□□□■■□■■■□□■□□■■	15	4
□□■■□■■■□□■□□■■□□□■■□■■□□□■□□■■	15	12
■□□■■■□■■□□□■■□□■□□■■■□□■□□□■■□	15	13
■■□■■□□■■■□□■□□□■■□■■□□□■■□□■□□	15	15
□■■■□□■■□■■□□□■□□■■■□□■□□■■□□□■	15	14
□■■□□□■■□■■■□□■□□■■□□□■□□■■■□□■	15	10
■■□□■□□■■■□■■□□□■■□□■□□□■■□■■□□	15	11
■□□□■■□■■□□■■■□□■□□□■■□□■□□■■■□	15	9
□□■□□■■■□□■■□■■□□□■□□■■□□□■■□■■	15	8
□□■□□■■■□□■■□■■■□□■□□■■□□□■■□■■	16	24
■□□□■■□■■□□■■■□■■□□□■■□□■□□■■■□	16	25
■■□□■□□■■■□■■□□■■■□□■□□□■■□■■□□	16	27
□■■□□□■■□■■■□□■■□■■□□□■□□■■■□□■	16	26
□■■■□□■■□■■□□□■■□■■■□□■□□■■□□□■	16	30
■■□■■□□■■■□□■□□■■■□■■□□□■■□□■□□	16	31
■□□■■■□■■□□□■■□■■□□■■■□□■□□□■■□	16	29
□□■■□■■■□□■□□■■■□□■■□■■□□□■□□■■	16	28
□□■■□■■□□□■□□■■■□□■■□■■■□□■□□■■	16	20
■□□■■■□□■□□□■■□■■□□■■■□■■□□□■■□	16	21
■■□■■□□□■■□□■□□■■■□■■□□■■■□□■□□	16	23
□■■■□□■□□■■□□□■■□■■■□□■■□■■□□□■	16	22
□■■□□□■□□■■■□□■■□■■□□□■■□■■■□□■	16	18
■■□□■□□□■■□■■□□■■■□□■□□■■■□■■□□	16	19
■□□□■■□□■□□■■■□■■□□□■■□■■□□■■■□	16	17
□□■□□■■□□□■■□■■■□□■□□■■■□□■■□■■	16	16

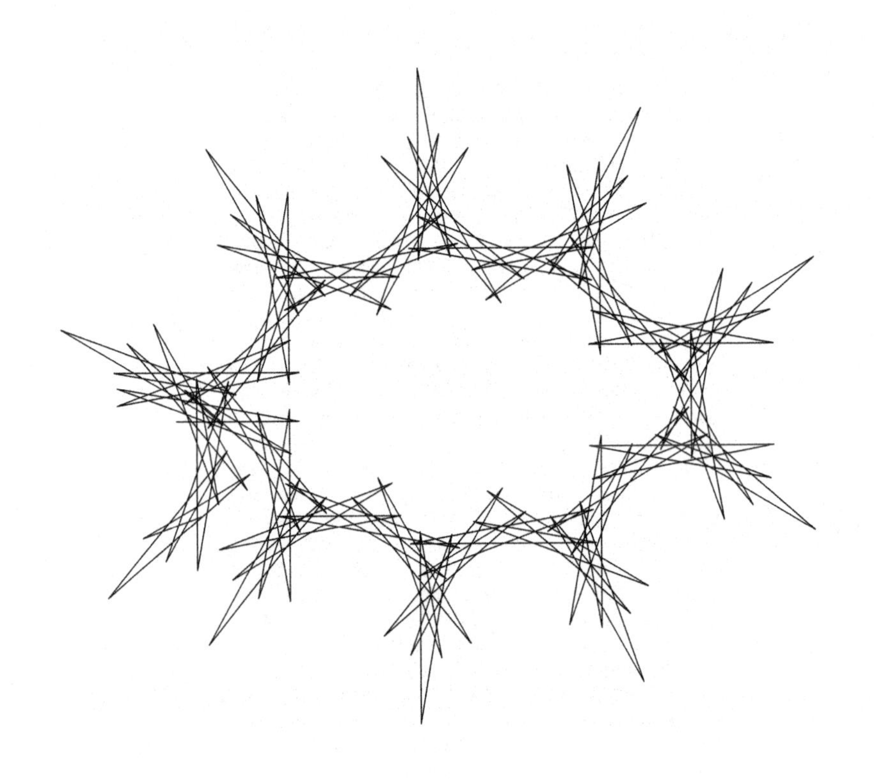

STOCHASTIC RHYTHMS

A series of beats spaced at a constant interval, like the ticking of a clock, is not by itself very interesting. Beats spaced at completely random intervals, such as rain or hail hitting the roof of a car, are also somewhat boring. The interesting rhythms are presumably somewhere in between these extremes.

We can explore this spectrum of rhythms, from the constant to the completely random, by generating random numbers with different degrees of correlation. Correlation is a measure of how well the next beat can be predicted from the last. The constant interval rhythm corresponds to 100% correlation and the completely random rhythm to 0% correlation.

The random numbers we are going to generate will represent the intervals between note onsets or beats. We will show how to control the strength of the correlation (predictability) of the intervals. Start by picking a maximum interval length. We'll use the number 8 as an example. How do we randomly generate intervals of length up to size 8? One way to do it is to toss 8 coins and count the number of heads. If you do this repeatedly, you will generate a random sequence of numbers that range from 0 to 8.

The probability distribution for these numbers is called the binomial distribution. A probability distribution simply tells you what the probability is for getting a particular number. The binomial distribution for the case where we have 8 coins and they are all assumed to be fair, is:

$$P(k) = \binom{8}{k}\frac{1}{2^8}$$

And for reference, the binomial distribution for the general case of n coins, where each coin has a probability p of coming up heads, is

$$P(n,k) = \binom{n}{k}p^k(1-p)^{n-k}$$

The probabilities for our particular case are listed in the table below. The most probable number is 4, which

k	$P(k)$
0	0.00390625
1	0.03125
2	0.109375
3	0.21875
4	0.2734375
5	0.21875
6	0.109375
7	0.03125
8	0.00390625

Table 36: Binomial distribution for 8 fair coins.

will occur about 27% of the time. The least probable

numbers are 0 and 8. The next most probable numbers are 1 and 7, and so on.

Not all the numbers have the same probability, but this does not mean that they are correlated. Every toss of the 8 coins is completely independent of the previous toss. The sequence of numbers produced by this process is completely random and uncorrelated. Suppose for example, a toss produces the number 4. This does not affect the possible values of the next toss in any way. The possible values still range from 0 to 8 with probabilities given by table 36.

The way to introduce correlation into the sequence is not to retoss all 8 of the coins, but to pick only a certain number of them at random, and retoss only those. Let's look at what happens if we only pick 7 of the coins to retoss. The value of the retoss will now be a random number ranging from 0 to 7 that also has a binomial distribution. This number added to the value of the coin that was not tossed will produce the next number in the sequence. The probabilities of the next number will now depend on the value of the current number. For example if the current number is 0 then the next number can not be 8 since the value of the untossed coin must be 0 and the value of the retossed coins cannot be larger than 7. This process will produce numbers that have the weakest possible correlation.

The strongest correlation is of course where we don't retoss any of the coins so the number just constantly repeats. The next strongest correlation occurs when only one of the coins is retossed. The value of the next number will then either equal the current number or be one less than or one greater than the current number. So the idea is that the correlation can be controlled by choosing how many coins to retoss at each step. Not retossing any will just repeat the same number continuously, resulting in 100% correlation. Retossing all of them will produce a random sequence with 0% correlation. The more coins retossed, the weaker the correlation.

The program `rndint.c` will produce random numbers using the process just described. You tell it the range of numbers you want it to produce, the starting number, how many of the coins to retoss at each step, and how many numbers to generate. The random numbers can be used as intervals for rhythms. Below are example outputs for the range 0 to 8, starting number 4, and for retossing 1, 4, and 8 coins.

```
rndint 8 4 1 64
4 4 5 5 6 5 5 4 4 4 3 2 2 1 1 1
0 0 0 0 0 0 0 1 1 1 1 2 2 3 2 3
2 1 1 1 2 2 2 3 2 2 3 4 3 4 4 5
5 5 6 5 5 5 6 7 6 5 5 5 4 5 5 6
```

```
rndint 8 4 4 64
```

```
4 3 3 4 6 7 7 6 7 7 6 5 6 5 3 4
5 3 3 3 4 4 4 5 2 5 5 6 7 7 5 4
5 5 4 2 5 5 5 5 2 2 4 4 5 3 3 4
2 3 4 3 3 2 2 4 5 6 5 5 4 3 4 5

rndint 8 4 8 64
4 3 2 5 6 3 3 5 3 4 4 3 3 4 6 4
5 2 7 5 4 2 5 3 4 5 1 1 6 3 3 3
4 2 6 3 5 5 3 4 1 5 4 3 5 3 6 4
5 4 2 5 4 4 5 5 3 3 3 4 4 4 5 4
```

Next we will look a little more closely at the mathematics behind this method of generating rhythms. It is for the more mathematically inclined reader who wants to explore these ideas further. If this is not you, then feel free to skip it.

The stochastic process we described above for generating random numbers can best be described as a Markov chain. The chain has $n+1$ states labeled from 0 to n where n is the number of coins used. The chain starts in some state and at each step a transition is made to a new state, or possibly the same state, with a probability that depends on the current state. The output of the chain is the series of states visited which are given by the numbers 0 to n. The properties of the sequence of numbers produced will depend on the transition probabilities.

To find the transition probabilities let's start by looking

at the mechanics of using n coins to make the transition. Suppose first of all that the chain is in state m. This means that m of the n coins are heads (value=1) and $n-m$ of them are tails (value=0). To get to the next state, k of the n coins are selected at random and tossed. If s is the number of heads on the selected coins before they are tossed and r is the number of heads after they are tossed, then the net change in heads is $r-s$ so the next state will be $\acute{m} = m + r - s$.

To find the transition probability to state $\acute{m}$ you need the probabilities for all the values of r and s such that $\acute{m} = m + r - s$. To get the probabilities for s, note that k coins are selected from a population of n coins of which m are heads and $n-m$ are tails. The probability of getting s heads in this selection is given by the hypergeometric probability distribution:

$$P(s) = \frac{\binom{m}{s}\binom{n-m}{k-s}}{\binom{n}{k}}$$

The probabilities for r are given by the binomial distribution for tossing k coins:

$$P(r) = \binom{k}{r}\frac{1}{2^k}$$

The transition probability from state m to state $\acute{m}$ is found by summing the product $P(s)P(r)$ for all values of s and r such that $\acute{m} = m + r - s$.

The transition probabilities are conditional probabilities which we will denote as $P(\acute{m}|m)$. They are usually collected together and used in a matrix called a transition matrix or stochastic matrix which we will denote as M. The element of M in row i and column j is the probability of moving to state j given that the current state is i, i.e. $M_{i,j} = P(j|i)$. The transition matrices for $n = 4$ and $k = 1, 2, 3$ are shown below.

$$\begin{pmatrix} \frac{1}{2} & \frac{1}{2} & 0 & 0 & 0 \\ \frac{1}{8} & \frac{1}{2} & \frac{3}{8} & 0 & 0 \\ 0 & \frac{1}{4} & \frac{1}{2} & \frac{1}{4} & 0 \\ 0 & 0 & \frac{3}{8} & \frac{1}{2} & \frac{1}{8} \\ 0 & 0 & 0 & \frac{1}{2} & \frac{1}{2} \end{pmatrix}$$

$$\begin{pmatrix} \frac{1}{4} & \frac{1}{2} & \frac{1}{4} & 0 & 0 \\ \frac{1}{8} & \frac{3}{8} & \frac{3}{8} & \frac{1}{8} & 0 \\ \frac{1}{24} & \frac{1}{4} & \frac{5}{12} & \frac{1}{4} & \frac{1}{24} \\ 0 & \frac{1}{8} & \frac{3}{8} & \frac{3}{8} & \frac{1}{8} \\ 0 & 0 & \frac{1}{4} & \frac{1}{2} & \frac{1}{4} \end{pmatrix}$$

$$\begin{pmatrix} \frac{1}{8} & \frac{3}{8} & \frac{3}{8} & \frac{1}{8} & 0 \\ \frac{3}{32} & \frac{5}{16} & \frac{3}{8} & \frac{3}{16} & \frac{1}{32} \\ \frac{1}{16} & \frac{1}{4} & \frac{3}{8} & \frac{1}{4} & \frac{1}{16} \\ \frac{1}{32} & \frac{3}{16} & \frac{3}{8} & \frac{5}{16} & \frac{3}{32} \\ 0 & \frac{1}{8} & \frac{3}{8} & \frac{3}{8} & \frac{1}{8} \end{pmatrix}$$

We will end by noting that these are not the only possible transition matrices. The only requirement for a transition matrix is that the probabilities in each row must sum to 1. You could for example have a matrix for a chain where state 1 is always followed by a transition to state 4. The matrix in this case would have zeros in all columns of row 1 except column 4 which would equal 1. This would produce rhythms where an interval of length 1 is always followed by an interval of length 4. The program `markovgen.c` will generate sequences using any transition matrices you give it. You can use it to custom design your own stochastic rhythms.

APPENDIX A: BOX NOTATION FOR SOME TRADITIONAL RHYTHMS

All the traditional rhythms below have midi files you can listen to on the book's website. Next to each rhythm is the name of the MIDI percussion instrument used. For more on traditional rhythms see Berry and Gianni.

Afro-Cuban

■□■□■■□■□■□■ Cowbell
□□□□■■□□□□□□ High Tom
□□■□□□□□■□□□ Electric Snare
□□□□□□□□□□■■ Low Tom
■□□□□□□□□□□■ Bass Drum 1
■□□■□□■□□■□□ Closed Hi Hat

Beguine

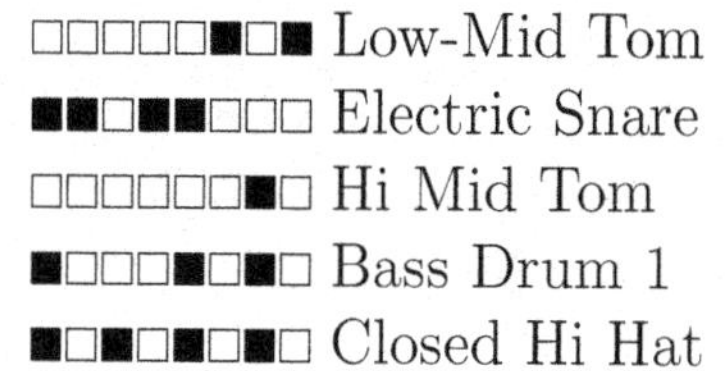

Bikutsi

■■■□■■■□■■■□ Closed Hi Hat
■□□□■□□□■□□□ Electric Snare
■□□■□□■□□■□□ Bass Drum 1

Bluegrass Train

■■■■■■■■■■■■■■■■ Electric Snare
■□□□■□□□■□□□■□□□ Bass Drum 1
□□■□□□■□□□■□□□■□ Closed Hi Hat

Blues Shuffle

■□■■□■■□■■□■ Closed Hi Hat
□□□■□□□□□■□□ Electric Snare
■□□□□□■□□□□□ Bass Drum 1

Bolero

■□■■■■□■□■□■□■□■ Closed Hi Hat
□□□□□□□■□■□□□■□□ Bass Drum 1

Bomba

■□■■□■■□■□■■□■■□ Cowbell
□■□□■□□■□■□□■□□■ Electric Snare
■□□■□□■□■□□■□□■□ Bass Drum 1
■□□□■□□□■□□□■□□□ Pedal Hi-Hat

Bossa Nova

■■■■■■■■■■■■■■■■ Ride Bell
■□□■□□■□□□■□□■□□ Electric Snare
■□□■■□□■■□□■■□□■ Bass Drum 1
□□■□□□■□□□■□□□■□ Closed Hi Hat

Cajun Zydeco

■□■□■□■□ Ride Cymbal 1
□□■□□□■□ Electric Snare
■□□■□□■□ Bass Drum 1
□□■□□□■□ Closed Hi Hat

Cascara

■□■■□■□■■□■□■■□■ Closed Hi Hat
■□□■□□■□□□■□■□□□ Electric Snare
□□□■□□■□□□□■□□■□ Bass Drum 1

Chacha

■□■□■□■□ Cowbell
□□□□□□■■ High Tom
□□■□□□□□ Electric Snare
□□□■■□□□ Bass Drum 1
■□■□■□■□ Closed Hi Hat

Conga

□□□□■■□□□□□□■□■□ Open Hi Conga
■□■□□□□■□■■□□□□□ Low Conga
□■□■□□■□■□□■□■□■ Electric Snare
■□□■□□■□■□□■□□■□ Bass Drum 1
■□□□■□□□■□□□■□□□ Closed Hi Hat

Cumbia

■□■■■□■■ Cowbell
□□■□□□■□ Electric Snare
■□□□■□■□ Bass Drum 1

Funk

■■■■■■■■ Electric Snare
■□□■□□■□ Bass Drum 1
□□■□□□■□ Closed Hi Hat

Gahu

■□□■□□■□□□■□□□■□ Low & High Tom

Heavy Metal

■■■■■■■■ Electric Snare
□□■□□□■□ Bass Drum 1
■□□□■□□□ Closed Hi Hat

Hip Hop

■■■■■□■■ Closed Hi Hat
□□■□□■■□ Electric Snare
■□□□□□■□ Bass Drum 1

Jazz

■■■■■■■■■■■■■■■■ Electric Snare
■□□□■□□□■□□□■□■□ Bass Drum 1
□□■□□□■□□□■□□□■□ Closed Hi Hat

Latin Rock

■■■■■■■■ Cowbell
□□□□□□■■ High Tom
□□■□□□□□ Electric Snare
■□□■■□□□ Bass Drum 1

Mambo

■□■□■■■■□■■■■□■■ Ride Bell
□□□□□□■■□□□□□□■■ High Tom
□□■□□□□□□□□□□□□□ Electric Snare
□□□□□□□□□□□■■□□□ Low Floor Tom
□□□■□□□□□□□■□□■□ Bass Drum 1
■□□□■□□□■□□□■□□□ Pedal Hi-Hat

Middle East

■□■■□■■□ Closed Hi Hat

□□□■□□■□ Electric Snare
■□□□□□□□ Bass Drum 1

Mozambique

■□■□■■□■□■■□■■□■ Cowbell
□□□□□□□□□□□■□□□□ High Tom
□□□□□□□■□□□□□□□□ Electric Snare
□□□□□□□□□□□□□□■□ High Floor Tom
□□□□■□□□□□□□□□□□ Bass Drum 1
■□□□■□□□■□□□■□□□ Pedal Hi-Hat

Palito

□□■□■□□□■□□■□□□■ Closed Hi Hat
■□□□□■□■□□■□□■□□ Electric Snare
□□□■□□■□□□□■□□■□ Bass Drum 1

Pilon

■□■■■□■■■□■■■□■■ Cowbell
□□□□□□□□□□□■□□□□ Low-Mid Tom
□□■□□□■□□□□□□□□□ Electric Snare
□□□□□□□□□□□□■□■■ Low Floor Tom
■□□□■□□□■□□□■□□□ Bass Drum 1
■□□□■□□□■□□□■□□□ Pedal Hi-Hat

Plena

□□■■□□■■□□■■□□■■ Cowbell
□□□□□□■□□□□□□□■□ Hi Mid Tom
□□■□□□□□□■□■□□□□ Electric Snare
■□□□■□□□■□□□■□□□ Bass Drum 1
■□□□■□□□■□□□■□□□ Pedal Hi-Hat

Polka

■■■■ Closed Hi Hat
□■□■ Electric Snare
■□■□ Bass Drum 1

Rai Maghreb

■□□■■□■□□■■□ Closed Hi Hat
■□■■□■■□■■□■ Electric Snare
□□□□■□□□□□■□ Bass Drum 1

Reggae

■□■■□■■□■■□■ Closed Hi Hat
□□□□□□■□□□□□ Electric Snare
□□□□□□■□□□□□ Bass Drum 1

Rock

■■■■■■■■ Closed Hi Hat
□□■□□□■□ Electric Snare
■□□□■□□□ Bass Drum 1

Rumba Clave

■□□■□□□■□□■□■□□□ Low & High Tom

Shiko

■□□□■□■□□□■□■□□□ Low & High Tom

Son Clave

■□□■□□■□□□■□■□□□ Low & High Tom

Songo

■□□□■□□□■□□□■□□□ Ride Bell

□□■□□■□■□■■□□■□■ Electric Snare
□□□■□□■□□□□■□□■□ Bass Drum 1

Soukous

■□□■□□■□□□■■□□□□ Low & High Tom

Surf

■■■■■■■■ Closed Hi Hat
□□■■□□■□ Electric Snare
■□□□■□□□ Bass Drum 1

Western

■□■■□■■□■■□■ Closed Hi Hat
□□□■□□□□□■□□ Electric Snare
■□□□□□■□□□□□ Bass Drum 1

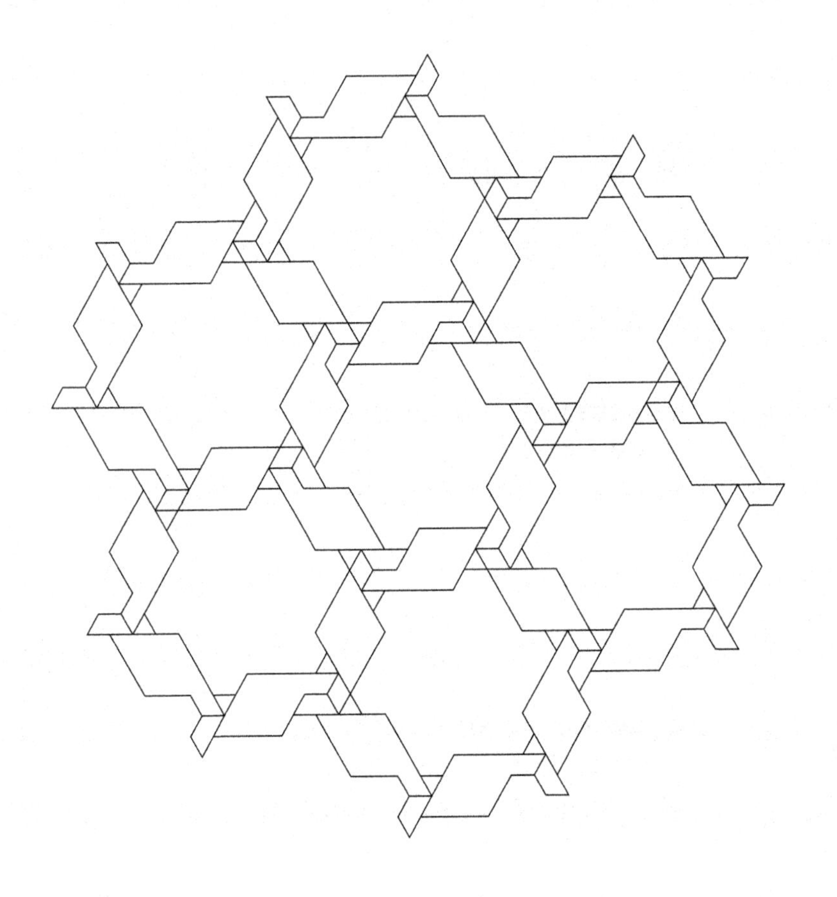

APPENDIX B: SOFTWARE

comp

```
Generates all compositions of n.

Usage: comp n
```

compm

```
Generates all compositions of n into m parts.

Usage: compm n m
```

compa

```
Generates all compositions of n with parts in
the set (p1 p2 ... pk).

Usage: compa n p1 p2 ... pk
```

compam

```
Generates all compositions of n with m parts
in the set (p1 p2 ... pk).

Usage: compam n m p1 p2 ... pk
```

comprnd

```
Generates a random composition of n.

Usage: comprnd n
```

compmrnd

Generates a random composition of n into m parts.

Usage: compmrnd n m

neck

Generates all binary necklaces of length n.

Usage: neck n

neckm

Generates all binary necklaces of length n with m ones.

Usage: neckm n m

necka

Generates all binary necklaces of length n with parts in (p1 p2 ... pk). A part is the length of a substring 10...0 composing the necklace. For example the necklace 10100 has parts of size 2 and 3.

Usage: necka n p1 p2 ... pk

neckam

Generates all binary necklaces of length n with m ones and parts in (p1 p2 ... pk). For a definition of parts see necka description.

Usage: neckam n m p1 p2 ... pk

part

Generates all partitions of n.

Usage: part n

partm

Generates all partitions of n into m parts.

Usage: partm n m

parta

Generates all partitions of n with parts in the set (p1 p2 ... pk).

Usage: parta n p1 p2 ... pk

partam

Generates all partitions of n with m parts in the set (p1 p2 ... pk).

Usage: partam n p1 p2 ... pk

permi

Generates all permutations of the non-negative integers in the set (a1 a2 ... an). To generate all permutations the integers must be ordered: a1 < a2 < ... < an. Any other order will only generate permutations larger in lexicographic order.

Usage: permi a1 a2 ... an

debruijn

```
Generates the largest de Bruijn sequence of
order n.

Usage: debruijn n
```

b2int

```
Reads binary strings from stdin and converts
them to interval notation.

Usage: b2int

Example:
  echo "1010010001001000" | b2int -> 2 3 4 3 4
```

int2b

```
Reads intervals from stdin and converts them
to binary string notation.

Usage: int2b

Example:
  echo "2 3 4 3 4" | int2b -> 1010010001001000
```

chsequl

```
Generates the upper or lower Christoffel word
for p/q.

Usage: chsequl t p q n
  t = type of word
    u = upper
    l = lower
  p = numerator
  q = denominator
  n = number of terms to generate, default=p+q
```

cfsqrt

```
Calculates the continued fraction for the
square root of an integer.

Usage: cfsqrt n
  n = integer
```

cfcv

```
Calculates a continued fraction convergent.

Usage: cfcv a0 a1 a2 ... an
  ai = simple continued fraction term
```

pfold

```
Generates fold sequences.

Usage: pfold n m f
  n = number of terms, 1,3,7,15,31,63,127,...
  m = number of bits
  f = function number 0 -> 2^m-1
```

rndint

```
Generates random numbers with specified
correlation.

Usage: rndint m s c n
  m = range of numbers, 0 to m
  s = starting number, 0 to m
  c = degree of correlation
      0 = total correlation (all numbers = s)
      m = no correlation (each number is
            independent)
  n = how many random numbers to generate
```

markovgen

```
Generates random numbers using a Markov chain.

Usage: markovgen mfile s n
  mfile = transition matrix file name
  s = starting state
  n = how many random numbers to generate
```

FURTHER READING

- *The Geometry of Musical Rhythm: What Makes a Good Rhythm Good?*, Godfried T. Toussaint, 2013

- *The Drum: A History*, Matt Dean, 2011

- *The Drummer's Bible: How to Play Every Drum Style from Afro-Cuban to Zydeco*, Berry and Gianni, 2012

- *Automatic Sequences: Theory, Applications, Generalizations*, Allouche and Shallit, 2003

- *Combinatorics on Words: Christoffel Words and Repetitions in Words (CRM Monograph)*, Berstel and Lauve and Reutenauer and Saliola, 2009

- *Finite Automata and Regular Expressions: Problems and Solutions*, Hollos and Hollos, 2013

- *Rhythm Wikipedia page*

- *Clave (rhythm) Wikipedia page*

- *Drum Wikipedia page*

- *General MIDI Wikipedia page*

ACKNOWLEDGMENTS

In ordinary life we hardly realize that we receive a great deal more than we give, and that it is only with gratitude that life becomes rich. It is very easy to overestimate the importance of our own achievements in comparison with what we owe to others.

Dietrich Bonhoeffer, letter to parents from prison, Sept. 13, 1943

We'd like to thank our parents, Istvan and Anna Hollos, for helping us in many ways.

We thank the makers and maintainers of all the software we've used in the production of this book, including: the Emacs text editor, the LaTeX typesetting system, LaTeXML, Inkscape, Evince document viewer, Maxima computer algebra system, gcc, awk, sed, bash shell, and the GNU/Linux operating system.

ABOUT THE AUTHORS

Stefan Hollos and **J. Richard Hollos** are physicists by training, and enjoy anything related to math, physics, and computing. They are the authors of

- **Pattern Generation for Computational Art**
- **Finite Automata and Regular Expressions: Problems and Solutions**
- **Probability Problems and Solutions**
- **Combinatorics Problems and Solutions**
- **The Coin Toss: Probabilities and Patterns**
- **Pairs Trading: A Bayesian Example**
- **Simple Trading Strategies That Work**
- **Bet Smart: The Kelly System for Gambling and Investing**
- **The QuantWolf Guide to Calculating Bond Default Probabilities**
- **The Mathematics of Lotteries: How to Calculate the Odds**
- **Signals from the Subatomic World: How to Build a Proton Precession Magnetometer**

They are brothers and business partners at Exstrom Laboratories LLC in Longmont, Colorado. Their website is exstrom.com